Activities Workbook

JAY WITHGOTT • SCOTT BRENNAN

CONTRIBUTORS

DAWN KELLER
Hawkeye Community College

MAY BOEVE
Co-Coordinator, 350.org

Benjamin Cummings
is an imprint of

San Francisco   Boston   New York
CapeTown   Hong Kong   London   Madrid   Mexico City
Montreal   Munich   Paris   Singapore   Sydney   Tokyo   Toronto

Senior Acquisitions Editor: Chalon Bridges
Project Manager, Editorial: Tim Flem
Sponsoring Editor: Amanda Brown
Managing Editor, Science: Gina M. Cheselka
Supplement Cover Designer: EW Design
Senior Operations Specialist: Alan Fischer
Cover Credits: Paul Nicklen/National Geographic/Getty (left); Bill Hatcher/National Geographic/Getty Images (right)

ISBN-10: 0-321-59057-0
ISBN-13: 978-0-321-59057-2

Copyright ©2009 Pearson Education, Inc., publishing as Benjamin Cummings, 1301 Sansome St., San Francisco, CA 94111. All rights reserved. Manufactured in the United States of America. This publication is protected by Copyright and permission should be obtained from the publisher prior to any prohibited reproduction, storage in a retrieval system, or transmission in any form or by any means, electronic, mechanical, photocopying, recording, or likewise. To obtain permission(s) to use material from this work, please submit a written request to Pearson Education, Inc., Permission Department, 1900 E. Lake Ave., Glenview, IL 60025. For information regarding permissions, call (847) 486-2635.

Many of the designations used by manufacturers and sellers to distinguish their products are claimed as trademarks. Where those designations appear in this book, and the publisher was aware of a trademark claim, the designations have been printed in initial caps or all caps.

Printing; 5 6 7 8

Benjamin Cummings
is an imprint of

Table of Contents

Introduction		v
Topic 1	Ecological Footprints	1
Topic 2	The Economy and the Environment	9
Topic 3	Environmental Policy	15
Topic 4	Chemistry and the Environment	21
Topic 5	Evolution and Biodiversity	27
Topic 6	Community Ecology	33
Topic 7	Ecosystem Ecology	41
Topic 8	Human Population	47
Topic 9	Soils	53
Topic 10	Agriculture and the Food We Eat	59
Topic 11	The Importance of Protecting Biodiversity	65
Topic 12	Forest Management	71
Topic 13	Urban Land Use	77
Topic 14	Environmental Health and Toxicology	83
Topic 15	Freshwater Resources	89
Topic 16	Marine Resources	95
Topic 17	Air Pollution and the Atmosphere	101
Topic 18	Global Climate Change	107
Topic 19	Fossil Fuels	113
Topic 20	Conventional Energy Alternatives	119

Topic 21	New Renewable Energy Alternatives	127
Topic 22	Waste Management	135
Topic 23	Sustainable Solutions	141
Topic 24	Geology, Minerals, and Mining	149

Introduction

As an Environmental Science instructor, I am always looking for the best materials for my students to use. That's why I adopted *Essential Environment: The Science behind the Stories* for my Environmental Science classes. It is well written, timely, and the right length to fit my needs. However, I was still struggling to find some sort of text to supplement my Environmental Science labs. That is why I was particularly excited to be a part of this new Activities Workbook.

The Activities Workbook pulls together various activities from the texts *Environment: The Science behind the Stories, 3rd edition*, and *Essential Environment: The Science behind the Stories, 3rd edition*, and provides space for students to write their answers. It is intended for use in a lecture or lab in which students are expected to interpret and analyze material and develop their own informed opinions about current environmental issues. The 24 topics follow the table of contents for *Environment: The Science behind the Stories, 3rd edition*, with the addition of a final topic on Geology, Minerals, and Mining (pertaining to new coverage in *Essential Environment, 3rd edition*). In addition, this workbook also includes new Case Studies written by May Boeve, whose contribution truly made this a more complete volume. I hope you find it to be a helpful addition to your classes.

Finally, I would like to thank Chalon Bridges, Senior Acquisitions Editor, for asking me to be a part of this project and Amanda Brown, Associate Editor, and Tim Flem, Project Manager, for always pointing me in the right direction.

Dawn Keller,
Hawkeye Community College
Waterloo, IA

Topic 1: Ecological Footprints

Causes and Consequences

CAUSES AND CONSEQUENCES

People in different societies consume resources and use land and water at different rates, according to their level of wealth or poverty, the technologies available to them, their society's cultural norms, and other factors. Societies whose people have **large ecological footprints** exert greater per capita impacts on Earth's natural resources and ecological systems. Fortunately, there are many solutions and steps that people can take to address the causes and mitigate the consequences of large ecological footprints.

Write in two causes of large ecological footprints in the spaces provided. Then write in two consequences (impacts on the environment, human health, or quality of life) that result from large ecological footprints. Finally, offer two solutions corresponding to this issue and its consequences. One cause, one consequence, and one solution have been filled in for you, providing examples.

CAUSES
- Strong demand for material goods
- _____

Large Ecological Footprints

CONSEQUENCES
- Depletion of natural resources
- _____

SOLUTIONS
- Improve efficiency of manufacturing processes
- _____
- _____

Sometimes solutions can have unintended consequences. Can you think of an undesired consequence that one solution to this issue might have? How might we then deal with *that* consequence?

Copyright © 2008 Pearson Education, Inc., publishing as Pearson Benjamin Cummings

Interpreting Graphs and Data

Environmental scientists study phenomena that range in size from individual molecules to the entire Earth and that occur over time periods lasting from fractions of a second to billions of years. To simultaneously and meaningfully represent data covering so many orders of magnitude, scientists have devised a variety of mathematical and graphical techniques, such as exponential notation and logarithmic scales. On the next page are two graphical representations *of the same data*, representing the growth of a hypothetical population from an initial size of 10 individuals at a rate of increase of approximately 2.3% per generation. The graph in part (a) uses a conventional linear scale for the population size; the graph in part (b) uses a logarithmic scale.

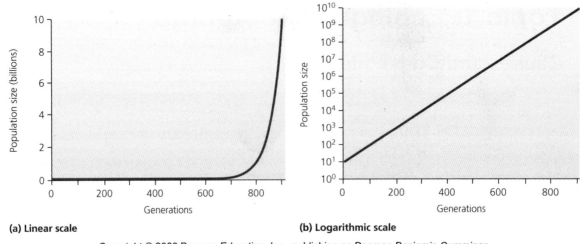

(a) Linear scale (b) Logarithmic scale

Hypothetical population growth curves, assuming an initial size of 10 and a constant rate of increase of approximately 2.3% per generation.

1. Using the graph in part (a), what would you say was the population size after 200 generations? After 600? After 800? After 900? How would you answer the same questions using the graph in part (b)?

2. What impression does the graph in part (a) give about population change for the first 600 generations? What impression does the graph in part (b) give? What impression does the graph in part (a) give about population change for the last 100 generations? What impression does the graph in part (b) give?

3. What advantages and disadvantages might there be in using a linear graph like the one in part (a)? What advantages and disadvantages might there be in using a logarithmic graph like the one in part (b)? What types of data would be clarified by each type of graph, and what types would be left unclear?

Calculating Ecological Footprints

Mathis Wackernagel and his many colleagues at the Global Footprint Network have continued to refine the method of calculating ecological footprints—the amount of biologically productive land and water required to produce the energy and natural resources we consume and to absorb the wastes we generate. According to their 2006 report, there are nearly 1.8 hectares (4.4 acres) available for every person in the world, yet we use on average more than 2.2 ha (5.4 acres) per person, creating a global ecological deficit, or overshoot of about 25%. Compare the ecological footprints of each of the countries listed in the table. Calculate their proportional relationships to the world population's average ecological footprint and to the land available globally to meet our ecological demands.

Country	Ecological footprint (hectares per person)	Proportion relative to world average footprint	Proportion relative to world area available
Bangladesh	0.5	0.224	
Colombia	1.3	.583	0.73 (1.3 ÷ 1.78)
Mexico	2.6	1.151	
Sweden	6.1	2.74	
Thailand	1.4	.628	
United States	9.6	4.305	
World average	2.23	1.0 (2.23 ÷ 2.23)	1.25 (2.23 ÷ 1.78)
Your personal footprint (see Question 3)			

Data from *Living planet report 2006*. WWF International, Zoological Society of London, and Global Footprint Network.

1. Why do you think the ecological footprint for people in Bangladesh is so small? Why is it so large for people in the United States?

2. Based on the data in the table, how do you think average per capita income affects ecological footprints? Explain your answer.

3. Go to an online footprint calculator such as the one at *www.myfootprint.org* or *www.ecofoot.org*, and take the test to determine your own personal ecological footprint. Enter the value you obtain in the table, and calculate the other values as you did for each nation. How does your footprint compare to those of people in the United States and in other nations? Name three actions you could take to reduce your footprint. (*Note:* Save this number—you will calculate your footprint again at the end of your course!)

Case Study

Global Footprint Network
Oakland, CA

Global Footprint Network is a research oriented non-profit organization that works with international partners in completing ecological footprinting. Mathis Wackernagel, who pioneered the Ecological Footprint concept, founded Global Footprint Network and today serves as the organization's Executive Director. By working globally, the organization hopes to popularize the concept and the usage of the Ecological Footprint as a system of accounting. One day, they hope to make the Ecological Footprint as widely known a measure of a country as its Gross Domestic Product. An additional goal, summed up by the eponymous name Ten-In-Ten, is for ten countries to manage their ecological resources as carefully and completely as they do their finances. They hope to achieve this by 2015.

One of the capstone concepts of the Ecological Footprint is "ecological overshoot," and it amounts to liquidating the planet's ecological resources and exceeding carrying capacity. The planet is already in overshoot, and the Global Footprint Network calculates the date each year when the planet has already exceeded its resources for the entire year. October 6 was so-called "ecological debt day" in 2007, and the date falls earlier each year. According to Global Footprint Network, this is a major threat, but not widely known. Today, the Ecological Footprint worldwide is over 23% larger than what the planet can regenerate. This means that it takes more than one year and two months for the Earth to regenerate what humans use in a single year.

To achieve its goals, the Global Footprint Network works with hundreds of partner organizations worldwide. They consult on and publish reports for countries, companies, and NGOs, and they collect and maintain the National Footprint Accounts, which measure resource usage and capacity of nations over time. Based on thousands of data points per country per year, the accounts calculate the footprints of 152 countries from 1961 to the present day. Any footprint calculated anywhere in the world can make use of this data.

Resources:

Global Footprint Network: *www.footprintnetwork.org*

Activity:

Given what you've learned about the ecological footprint, work together in small groups to determine your classroom's ecological footprint. Try to focus on the room itself, but include what resources are needed to power the classroom from outside of it.

Questions:

1. Given the emphasis on global climate change within environmental news, does the ecological footprint seem like an outdated concept?

2. How might you reduce your own personal ecological footprint?

3. Do you think the ecological footprint concept is an effective way to motivate people to care about environmental problems? Why or why not?

Topic 2: The Economy and the Environment

Interpreting Graphs and Data

As described in the text, economists are beginning to use various indicators of economic well-being as alternatives to the Gross Domestic Product (GDP), the total monetary value of final goods and services produced each year. An alternative measure called the Genuine Progress Indicator (GPI) is calculated as follows:

GPI = (GDP) + (Benefits ignored by GDP) - (Environmental costs) - (Social and economic costs)

Benefits include such things as the value of parenting and volunteer work. Environmental costs include the costs of water, air, and noise pollution; loss of wetlands; and depletion of nonrenewable resources, among others. Social and economic costs include investment, lending, and borrowing costs, as well as the costs of crime, family breakdown, underemployment, commuting, pollution abatement, automobile accidents, and loss of leisure time.

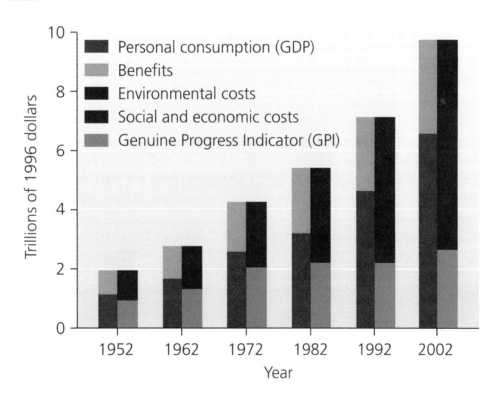

Components of GPI and GDP for the United States, 1952–2002. Data from Venetoulis, J., and C. Cobb. 2004. *The genuine progress indicator, 1950–2002 (2004 update)*. Redefining Progress.

1. Describe economic growth as measured by GDP for the United States from 1952 to 2002. Now describe economic growth as measured by GPI over the same time period. To what factors would you attribute the growing difference between these measures?

2. For GPI to grow significantly, one or more things must happen: Either GDP must grow faster; benefits ignored by the GDP must grow faster; or social, economic, and environmental costs must shrink relative to the other terms. Which of these scenarios do you think is most likely? Which would you prefer? How do the data in the graph support your answer?

3. Despite policy enacted in recent decades to regulate air and water pollution, hazardous waste disposal, solid waste management, forestry practices, and species protection, environmental costs continue to increase. Why do you suppose the trend remains in that direction?

Calculating Ecological Footprints

Although the Gross Domestic Product (GDP) of the United States grew impressively between 1952 and 2002 (see the graph in the "Interpreting Graphs and Data" section), so did the U.S. population. According to the U.S. Census Bureau, the midyear population of the nation was 157,552,740 in 1952, and 288,368,698 in 2002—an 83% increase. Estimate the values of the components of the Genuine Progress Indicator (GPI) for the United States in 1952 and 2002 from the graph in the previous section, and enter your estimates into the table on the next page. Then, using the population figures, calculate the per capita values for each component in 1952 and 2002.

Components of GPI	U.S. total in 1952 (trillions of dollars)	Per capita in 1952 (thousands of dollars)	U.S. total in 2002 (trillions of dollars)	Per capita in 2002 (thousands of dollars)
GDP	1.2	7.6	6.6	22.9
Benefits				
Environmental costs				
Social and economic costs				
GPI				

1. Consider your own life. What would you estimate is the value of the benefits in which you participate? Compare your personal estimate with the national average value in 2002.

2. What would you estimate are the values of the environmental, social, and economic costs for which you are personally responsible? How do they compare with the national average values in 2002?

3. In 2002, social and economic costs were proportionally larger relative to GDP than they were in 1952, and environmental costs were roughly the same in proportion to GDP. How would you account for these trends? What could you do to help improve these trends in your own personal accounting?

Case Study

Socially Responsible Investing at Williams College

Williams College
Williamstown, MA
Founded 1793
Undergraduate enrollment: 2,000

Socially Responsible Investing (SRI) is an increasingly popular practice that applies principles of social consciousness to financial investment. It represents an approach that influences private sector practices from within, rather than through policy campaigns or grassroots activism. Today SRI includes an estimated $2.71 trillion out of $25.1 trillion in the U.S. investment marketplace. SRI attempts to strike a compromise between the financial needs of investors and the societal results of investment decisions. The term "socially responsible" encompasses environmental, social, and governance issues. There are many varieties of socially responsible investors, including individuals and institutions. Institutional investors (universities, hospitals, and pension funds, for example) are the largest and fastest growing elements of the SRI sector.

One of the leading U.S. colleges pursuing SRI is Williams College, located in Williamstown, Massachusetts. According to the Sustainable Endowments Institute, an organization that independently ranks university environmental performance, Williams is one of a small number of leading educational institutions when it comes to SRI. Williams' endowment amounts to $1.9 billion, which is large for a college of its size. Here are some of the policies Williams has enacted to manage its endowment in a socially responsible fashion:

- Transparency: The college makes voting records and information about investment holdings available to the larger college community.
- Investment decisions: The college invests responsibly in areas including community development and clean energy. It also created a Social Choice Fund specifically for investors interested in these areas.
- Broader participation: The college created an Advisory Committee on Shareholder Responsibility that is open to student, faculty, alumni, and administrative participation.

Resources:

Social Investment Forum: *www.socialinvest.org/resources/sriguide/srifacts.cfm*

Sustainable Endowments Institute: *www.endowmentinstitute.org*

Williams College: *www.williams.edu*

Activity:

Organize into small groups and discuss the following questions: What do you think of socially responsible investing? How could you find out whether your college is participating in these types of investments? If your college is involved with SRI, do you think the public would be interested in knowing about it? If your college is not involved with SRI, how could you try to educate administrators about such programs?

Questions:

1. Do you think the general public needs to be made more aware of socially responsible investing? Explain your answer.

2. Prior to reading about this topic, did you think about your college's investment decisions? Has this topic changed your opinion about investing? Why or why not?

3. Should SRI become a priority for colleges and universities pursuing sustainability, or should their focus remain on issues such as recycling and climate change? Explain your answer.

Topic 3: Environmental Policy

Interpreting Graphs and Data

The Clean Air Act legislation of 1970, 1977, and 1990 was designed to improve air quality in the United States by monitoring and reducing the emissions of air pollutants judged to pose threats to human health, such as carbon monoxide, nitrogen dioxide, sulfur dioxide, ozone, particulate matter, and lead. The main source of lead emissions in 1970 was the exhaust of vehicles burning gasoline to which tetra-ethyl lead had been added to improve combustion. By 1985, leaded gasoline was phased out of use, although airplanes and racecars were exempted. The 1990 amendments addressed the growing problem of urban smog by requiring the use of reformulated gas (RFG) in cities with the worst smog problems. One of the RFG requirements specifies 2% oxygen content in fuel, which has been met by adding either ethanol or methyl tert-butyl ether (MTBE). Although it burns cleanly, MTBE is water-soluble and may cause cancer, so groundwater contamination from fuel spills is a concern.

Twenty-five states have now passed legislation banning or restricting the use of MTBE. The following graph shows trends in U.S. lead emissions and MTBE consumption since 1970.

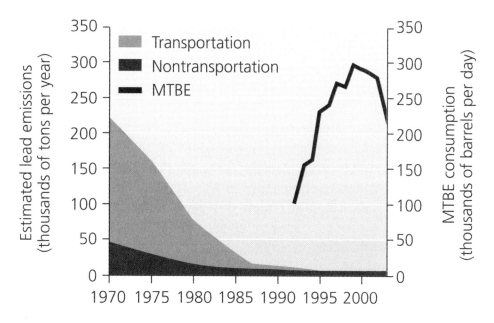

Estimated lead emissions from transportation and non-transportation sources, and consumption of MTBE in the United States. Data from U.S. Department of Transportation, Bureau of Transportation Statistics.

1. Did policy resulting from Clean Air Act legislation succeed in reducing the public health risk from exposure to lead? Use data provided in the graph to support your answer.

2. In the 1990s, use of MTBE in RFG increased rapidly on the East and West coasts of the United States. Ethanol blends were used mostly in the corn-growing states of the central part of the country; ethanol is made from corn, so it provides a market for in-state farmers while low transportation costs make it cheaper for in-state consumers. As MTBE use is being phased out in many states, use of ethanol is increasing. Name three ways in which these trends illustrate how state government actions may differ from federal government actions.

3. Do you think it is better policy to regulate air and water quality separately (e.g., under the Clean Air Act and Clean Water Act) or together under a single, comprehensive act to prevent and control pollution (as some European nations do)? How do the data in the graph support your reasoning?

Calculating Ecological Footprints

In 2000, the population of the United States was approximately 285 million. According to the U.S. Geological Survey, the country withdrew about 408 billion gallons of fresh water per day for a variety of uses, and average per capita use was 1,430 gallons per day. Seventy-nine percent of 2000 withdrawals were from surface water, and 21 percent of withdrawals were from groundwater. Thermoelectric power plants accounted for 48% of total withdrawals, irrigation 34%, public water supplies 11%, and industrial, livestock, and mining uses 8% (percentages add up to 101% because of rounding).

Daily Water Use in 2000				
	Electricity	Irrigation	Public water supply	Industrial / livestock / mining
You	686 gal			
Your class				
Your hometown				
United States				

Data from U.S. Geological Survey, *Summary of water use in the United States, 2000.*
http://ga.usgs.gov/edu/wateruse2000.html

1. How many gallons of the country's daily water use come from surface water?

2. What types of policies would you recommend to reduce overall water use most aggressively? Provide one example of a command-and-control policy and one example of an economic policy tool.

3. Describe how these recommended policies would affect withdrawals in one of the four water use categories listed on the previous page.

Case Study

Syracuse University
Syracuse, New York
Founded 1870
Undergraduate enrollment: 17,000

For some colleges and universities, investing in energy improvements is impossible without additional funding. In such cases, government grants may be made available. This can result in unique public-private partnerships. One example of this is an energy efficiency program at Syracuse University, funded in part by the New York State Energy Research and Development Authority (NYSERDA). NYSERDA provided a $552,000 grant to Syracuse University to help offset the cost of a $1.8 million investment into energy efficiency. Specifically, the grant was directed toward buying an energy-efficient chiller. The chiller will benefit the school economically and environmentally, saving $150,000 and 1.8 million kilowatt-hours of electricity each year.

NYSERDA fills an unusual niche in the public arena. In 1975, New York altered the mission of the State Atomic and Space Development Authority and transformed it into a research entity focused on energy. Specifically, NYSERDA's research aimed to reduce New York's petroleum consumption. Today, the mission has expanded to include research into new technologies, but still with a focus on energy.

These kinds of government and campus partnerships can be excellent sources of funding for campus sustainability projects. They also make policymaking more real to students, because the benefits of state policy accrue to their lives on campus.

Resources:

Syracuse University: *www.syracuse.edu*

NYSERDA: *www.nyserda.org*

Syracuse Solar and Wind: *www.syracusesolarandwind.com*

Activity:

Think of a particular need on your campus that could be met by a government-funded environmental sustainability initiative. How would you raise awareness about such a project? Who would you ask for help? How might your efforts help inform and benefit such initiatives at other schools?

Questions:

1. Is your campus involved in any efforts involving environmental sustainability? If so, what are they? If you aren't sure, how could you find out?

2. NYSERDA derives revenue from state rate-payers through a Systems Benefits Charge. If you were a New York state resident, would you be in favor of continuing the Systems Benefits Charge, and possibly expanding it, to improve energy efficiency programs?

3. Do you believe it is the government's responsibility to promote environmental sustainability through government-sponsored funding programs? Explain your answer.

Topic 4: Chemistry and the Environment

Interpreting Graphs and Data

In phytoremediation, plants are used to clean up soil or water contaminated by heavy metals such as lead (Pb), arsenic (As), zinc (Zn), and cadmium (Cd). For plants to absorb these metals from soil, the metals must be dissolved in soil water. For any given instance, all metal can be accounted for as either remaining bound to soil particles, being dissolved in soil water, or being stored in the plant. In a study on the effectiveness of alpine penny-cress (*Thlaspi caerulescens*) for phytoremediation, Enzo Lombi and his colleagues grew crops of this small perennial plant for approximately one year in pots of soil from contaminated sites. They then measured the amount of zinc and cadmium in the soil and in the plants when they were harvested.

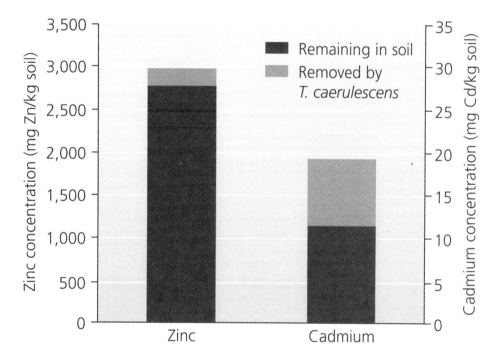

Removal of zinc and cadmium from contaminated soil by alpine penny-cress, *Thlaspi caerulescens*. Data from Lombi, E., et al. 2001. Phytoremediation of heavy metal-contaminated soils: Natural hyperaccumulation versus chemically enhanced phytoextraction. *Journal of Environmental Quality* 30:1919–1926.

1. What were the zinc and cadmium concentrations in the soil prior to phytoremediation? What were the zinc and cadmium concentrations in the soil after one year of phytoremediation?

2. How much zinc and cadmium were removed from the soil? If the plants continue to remove zinc and cadmium from the soil at the rates shown, how long would it take to remove all the zinc and cadmium?

3. Alpine penny-cress produces natural compounds that increase the solubility of metals in soil water. If these dissolved metals are not taken up by the plants, what may be an unintended consequence of having increased their solubility?

Calculating Ecological Footprints

In ecological systems, a rough rule of thumb is that when energy is transferred from plants to plant-eaters or from prey to predator, the efficiency is only about 10%. Much of this inefficiency is a consequence of the second law of thermodynamics. Another way to think of this is that eating 1 calorie of meat from an animal is the ecological equivalent of eating 10 calories of plant material. Humans are considered omnivores because we can eat both plants and animals. The choices we make about what to eat have significant ecological consequences. With this in mind, calculate the ecological energy requirements for four different diets, each of which provides a total of 2,000 dietary calories per day and write your answers in the table on the next page.

Diet	Source of calories	Number of calories consumed	Ecologically equivalent calories	Total ecologically equivalent calories
100% plant	Plan			
0% animal	Animal			
90% plant	Plant	1,800	1,800	3,800
10% animal	Animal	200	2,000	
50% plant	Plant			
50% animal	Animal			
0% plant	Plant			
100% animal	Animal			

1. How many ecologically equivalent calories would it take to support you for a year, for each of the four diets listed?

2. How does the ecological impact from a diet consisting strictly of animal products (e.g., milk, other dairy products, eggs, and meat) compare with that of a strictly vegetarian diet? How many additional ecologically equivalent calories do you consume each day by including as little as 10% of your calories from animal sources?

3. What percentages of the calories in your own diet do you think come from plant versus animal sources? Estimate the ecological impact of your diet, relative to a strictly vegetarian one.

4. Describe some challenges of providing food for the growing human population, especially as people in many poorer nations develop a taste for an American style diet rich in animal protein and fat.

Case Study

Bioremediation in New Orleans following Hurricane Katrina

After Hurricane Katrina struck the Gulf Coast, the crisis overtook the news cycle. Many people were outraged about the government's response to the hurricane, and many former residents wondered what might happen to their beloved city. Some of the problems associated with the hurricane were widely known: a dispersed population, extensive water and mold damage, and lack of safe housing for the people who remained in the city.

One of the lesser-known problems related to contaminated soil. Fortunately, as with the case of cleaning up the *Exxon-Valdez* oil spill, bioremediation may prove an excellent solution for New Orleans. A local organization, Common Ground, is advocating for this kind of long-term and ecologically sound solution. Focusing on one part of the city, Orleans Parish, Common Ground helped found the Meg Perry Community Garden and Bioremediation Project to work with local residents in addressing environmental hazards.

According to its website, since 2005 Common Ground's bioremediation volunteers have helped to remediate contaminated soil using natural methods, such as planting ferns in damaged soil so as to remove lead. Another aspect of Common Ground's program is growing mushrooms as a means of bioremediation. Mushrooms make good candidates for bioremediation because they can remove petrochemicals from soil, and following this, can be safely composted. The resulting compost binds metals to the soil, which in turn improves soil quality and productivity.

The program doesn't end there. Soil testing helps assure that no toxic chemicals remain in the soil following bioremediation. Volunteers collect soil samples, which are then tested at the Louisiana State University Agricultural Center. If the soil samples contain no toxic chemicals, the next step is determining which plants, if any, can be effectively grown in the soil. Some of the soil sample volunteers are local high school students.

Resources:

Meg Perry Community Garden: *www.commongroundrelief.org/bioremediation*

Activity:

Imagine you are planning a Spring Break trip to New Orleans to help with the cleanup efforts related to Hurricane Katrina. Bioremediation can be considered a long-term sustainability measure, as well as something to help with immediate clean-up needs. Do you think students would be interested in such a community service project? How would you encourage other students to get involved with such a project?

Questions:

1. Do you think bioremediation can be an effective means to educate people about environmental chemistry? How so? If not, what might be a better way?

2. Aside from large-scale cleanup projects, can you think of further applications for bioremediation? Describe what they might be.

3. List some potential limitations or problems you see with bioremediation.

Topic 5: Evolution and Biodiversity

Interpreting Graphs and Data

Amphibians are sensitive biological indicators of climate change because their reproduction and survival are so closely tied to water. One way in which drier conditions may affect amphibians is by reducing the depth of the pools of water in which their eggs develop. Shallower pools offer less protection from UV-B (ultraviolet-B) radiation, which some scientists maintain may kill embryos directly or make them more susceptible to disease. Herpetologist Joseph Kiesecker and colleagues conducted a field study of the relationships among water depth, UV-B radiation, and survivorship of western toad (*Bufo boreas*) embryos in the Pacific Northwest. In manipulative experiments, the researchers placed toad embryos in mesh enclosures at three different depths of water. The researchers placed protective filters that blocked all UV-B radiation over some of these embryos, while leaving other embryos unprotected without the filters. Some of the study's results are presented in the graph below.

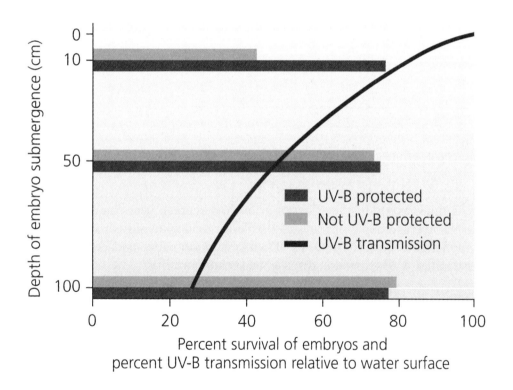

Embryo survivorship in western toads (*Bufo boreas*) at different water depths and UV-B light intensities. Red bars indicate embryos protected under a filter that blocked UV-B light; orange bars indicate unprotected embryos. The blue line indicates the amount of UV-B light reaching different depths in the water column, expressed as a percentage of the UV-B radiation at the water surface. Data from Kiesecker, J. M., et al. 2001. Complex causes of amphibian population declines. *Nature* 410: 681–684.

1. If the UV-B radiation at the surface has an intensity of 0.27 watts/m^2, approximately what is its intensity at depths of 10 cm, 50 cm, and 100 cm?

2. Approximately how much did survival rates at the 10-cm depth differ between the protected and unprotected treatments? Why do you think survival rates differed significantly at the 10-cm depth but not at the other depths?

3. What do you think would be the effect of drier-than average years on the western toad population, if the average depth of pools available for toad spawning dropped? How do the data in the graph address your hypothesis? Do they support cause-and-effect relationships among water depth, UV-B exposure, disease, and toad mortality?

Calculating Ecological Footprints

In 2004, coffee consumption in the United States topped 2.7 billion pounds (out of 14.8 billion pounds produced globally). Next to petroleum, coffee is the most valuable commodity on the world market, and the United States is its leading importer. Most coffee is produced in large tropical plantations, where coffee is the only tree species and is grown in full sun. However, approximately 2% of coffee is produced in small groves where coffee trees and other species are intermingled. These *shade-grown* coffee forests maintain greater habitat diversity for tropical rainforest wildlife. Given the information above, estimate the coffee consumption rates in the table below.

	Population	Pounds of coffee per day	Pounds of coffee per year
You (or the average American)	1	0.025	9
Your class			
Your hometown			
Your state			
United States			

Data from O'Brien, T. G., and M. F. Kinnaird. 2003. Caffeine and conservation. *Science* 300:587; and International Coffee Organization.

1. What percentage of global coffee production is consumed in the United States? If only shade-grown coffee were consumed in the United States, how much would shade-grown production need to increase to meet that demand?

2. How much extra would you be willing to pay for a pound of shade-grown coffee, if you knew that your money would help to prevent habitat loss or extinction for animals such as Sumatran tigers, rhinoceroses, and the many songbirds that migrate between Latin America and North America each year? Explain your answer.

3. If everyone in the United States were willing to pay as much extra per pound for shade-grown coffee as you are, how much additional money would that provide for conservation of biodiversity in the tropics each year?

Case Study

Lindblad Expeditions, the Galapagos Islands

The Galapagos Islands are known worldwide for their biodiversity. These islands and their species helped inform Charles Darwin's insights into the processes of evolution 150 years ago. Today, many thousands of tourists visit the Galapagos each year, hoping to have their own up-close encounters with the likes of Blue-footed Boobies and Marine Iguanas.

Most visitors to the Galapagos choose to visit with established ecotourism companies and their trained guides. One example is Lindblad Expedition's ecotourism program, which provides the opportunity to visit the Galapagos with a professional team of naturalists, and also offers swimming, snorkeling, and kayaking. This company is affiliated with *National Geographic*, which lends it increased credibility. The Lindblad family can claim a long history with the Galapagos, dating back to Eric Lindblad's first expedition in 1967. His trip marked the first non-scientific expedition to the Islands. Since that first trip, the family has continued its visits to the present day. Through these expeditions, the family's legacy now includes introducing more of the public to the biodiversity of the Galapagos.

The group also founded its own non-profit organization to support conservation in the Galapagos. Founded in 1997, the Galapagos Conservation Fund has raised over $3.6 million for its environmental efforts.

Lindblad Expeditions is one of many ecotourism agencies travelers can use to visit the Galapagos Islands, and most of these travel providers incorporate a conservation ethic into their work.

Resources:

Lindblad Expeditions: *www.expeditions.com/Destination44.asp?Destination=294*

Galapagos Conservancy: *www.galapagos.org*

Activity:

Imagine that an ecotourism agency moves into your campus community to guide trips to a nearby natural area. You apply for a summer job to develop its ecotourism program. How would you design the program? Consider the area your campus is in, and describe what appropriate ecotourism activities there might be in this region. Plan several trips on which you would guide people, and describe how you would recruit people to participate. Do you think this would be a lucrative endeavor?

Questions:

1. Do you think ecotourism is an example of a sustainable business? Explain your answer.

2. Have you ever participated in an ecotourism trip? Consider a broad definition of the term (this could include hiking, kayaking, or other forms of low-impact activity in the natural world). Do you need to be involved with an ecotourism company to say you are an ecotourist? Why or why not?

3. Ecotourism trips to the Galapagos Islands are popular for two reasons: (1) the islands hold fascinating biodiversity, and (2) the islands are famous for their historical significance in inspiring Charles Darwin's findings. Can you think of and name other places in the world that would make for good ecotourism ventures based on criteria like these?

Topic 6: Community Ecology

Causes and Consequences

CAUSES AND CONSEQUENCES

As globalization proceeds, as we become more mobile, and as our societies become more interconnected, people are transporting plants and animals among regions, nations, and continents. Of the many non-native species introduced, some become invasive and spread, displacing native flora and fauna. **Invasive species** are exerting ever-greater impacts on our natural communities and societies. Thankfully, we can take steps to prevent introductions, control the spread of invasives, and even eradicate certain exotic species.

Write in two causes of invasive species in the spaces provided. Then write in two consequences (impacts on the environment, human health, or quality of life) that result from invasive species. Finally, offer two solutions to this issue and its consequences. One cause, one consequence, and one solution have been filled in for you, providing examples.

CAUSES
Increased global transport and mobility

Invasive Species

CONSEQUENCES
Decline of native organisms from competition

SOLUTIONS
Policy aimed at ballast water dumping and other particulars

Sometimes solutions can have unintended consequences. Can you think of an undesired consequence that one solution to this issue might have? How might we then deal with *that* consequence?

Copyright © 2008 Pearson Education, Inc., publishing as Pearson Benjamin Cummings

Interpreting Graphs and Data

Each spring, citizen volunteers join representatives of the United States Geological Survey, the California Department of Fish and Game, and the Monterey Bay Aquarium to survey populations of the sea otter (*Enhydra lutris*) along 600 km (375 mi) of California's coastline, from Half Moon Bay to Santa Barbara. The information gathered from the survey is used by federal and state wildlife agencies in making decisions about how to manage the sea otter—a keystone species that in turn influences the coast's ecological communities.

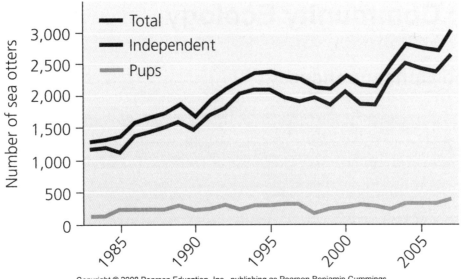

Numbers of sea otters counted during spring surveys, 1983–2007. Shown are numbers of independent adults and subadults, dependent pups, and total individuals. Data from USGS Western Ecological Research Center.

1. In percentage terms, how does the most recent population estimate of sea otters (from 2007) compare with the survey's earliest estimate (from 1983)?

2. Thinking about what you learned about keystone species (including the sea otter) from the text, describe how you think recent population trends for this species are likely to affect the composition and structure of the nearshore community in regard to kelp, sea urchins, and other organisms.

3. Given your hypotheses in Question 2, sketch three of your own graphs for the period 1983–2007 showing population trends for (a) sea urchins, (b) kelp, and (c) fish that depend on dense kelp forest habitat. Explain how your graphs illustrate predictions from your hypotheses.

Calculating Ecological Footprints

In 2003, scientist David Pimentel reviewed estimates for the economic and ecological costs of certain introduced and invasive species. He noted that over 50,000 species have been introduced in the United States (see Table 1) and estimated that damages and costs associated with controlling aquatic invasive species in the United States total $9 billion annually (see Table 2 on the next page). Calculate values missing from these tables.

Table 1. Introduced Species

Taxon	Percentage of total introduced	Number of species introduced
Plants	50	25,000
Mammals	0.04	
Birds	0.19	
Molluscs	0.18	
Arthropods	9	
Microbes	40	

Table 2. Aquatic Invaders

Invader	Percentage of total costs	Economic cost
Fish species	60	$5.4 billion
Zebra/quagga mussels	11	
Asiatic clam	11	
West Nile virus	11	
Aquatic plant species	5	
Shipworm	2	
Green crab	1	

Source: Pimentel, D., 2003. Economic and ecological costs associated with aquatic invasive species. *Proceedings of the Aquatic Invaders of the Delaware Estuary Symposium,* Malvern, Pennsylvania, May 20, 2003.

Copyright © 2008 Pearson Education, Inc., publishing as Pearson Benjamin Cummings

1. Of the over 50,000 species introduced into the United States, approximately half are plants. Describe how non-native plants may sometimes turn invasive and pose threats to native communities.

2. Compare the costs from aquatic invaders shown in Table 2 with the percentages shown for all non-native species in Table 1. Speculate on the costs you might expect from terrestrial introduced species, and explain your answers.

3. How might your own behavior influence the influx and ecological impacts of non-native species like those listed above? What could you do to minimize the impacts of invasive species?

Case Study

Point Reyes National Seashore
Marin County, California
Founded: 1962
Yearly visitors: 2.5 million

Non-native plants infest about 2.6 million acres in the U.S. National Parks system. To bring that number closer to home, consider the Point Reyes National Seashore north of San Francisco. In that park, 1/3 of all plant species are non-native. 30 of those non-native plants are especially invasive and present a more serious threat to native plant species throughout the park. In addition to the already threatened native plant species, 46 plant species are categorized as rare.

To place these numbers in context, the California Floristic Province (CFP), which Point Reyes is a part of, is one of only 25 terrestrial regions considered a global biodiversity hotspot. This designation, on behalf of Population Action International and the Nature Conservancy, relates to levels of biodiversity and high chances of species loss. In the CFP, less than 10% of these highly valued species are protected.

In order to help curb the problem of invasive non-native species, the Seashore initiated a program for volunteers to help uproot the problem plants. One particularly pesky plant is the Iceplant, which grows so thickly it creates a veritable carpet across coastal dunes and makes it impossible for native plants to seed, germinate, and thrive. Weed pulling is not everyone's idea of a good time, especially since so many of us have to do in our *own* yards, let alone at a national seashore. But nevertheless, the program is effective, and many weekends of the year volunteers can be seen crouched on the coastal dunes, methodically pulling up non-native plants.

Resources:

Point Reyes National Seashore: *www.nps.gov/pore/*

Activity:

Many biologically diverse regions are threatened by development, but not all of them appear to be worth protecting. The Point Reyes National Seashore, because it is protected, is a favorite tourist location for its natural beauty. But consider the example of wetlands--sometimes they look like nothing more than mud puddles, but are instead great sources of biodiversity. How would you convince a developer to not build on a wetland, or another unsightly but important area? Develop a list of five reasons to protect biodiversity. If you want to go an extra step, research local initiatives in your campus community to place new buildings in natural areas. Is there a community group opposed to development, and would your biodiversity recommendations be helpful to them?

Questions:

1. What are examples of non-native and native plant species in your area?

2. Depending on how many examples you think of for Question 1, how were you educated about these species? Were you able to think of one type of species more easily than the other? Do you think this information is widely known?

3. Does your campus implement landscaping plans integrating native species? Why or why not?

Topic 7: Ecosystem Ecology

Causes and Consequences

CAUSES AND CONSEQUENCES

The hypoxic zone in the Gulf of Mexico is not unique; at least 200 **dead zones** exist at estuaries and along seacoasts worldwide. These low-oxygen regions share many of the same causes, and they all have ecological and economic repercussions for ecosystems and people. Fortunately, we can pursue strategies to lessen the number, severity, and impacts of dead zones.

Write in two causes of dead zones in the spaces provided. Then write in two consequences (impacts on the environment, human health, or quality of life) that result from dead zones. Finally, offer two solutions to this issue and its consequences. One cause, one consequence, and one solution have been filled in for you, providing examples.

CAUSES
Nutrient runoff from agricultural fertilizers
_____
_____

Dead Zones

CONSEQUENCES
Death of aquatic organisms that cannot escape
_____
_____

SOLUTIONS
Steps to decrease nutrient pollution from farms
_____
_____

Sometimes solutions can have unintended consequences. Can you think of an undesired consequence that one solution to this issue might have? How might we then deal with *that* consequence?

Copyright © 2008 Pearson Education, Inc., publishing as Pearson Benjamin Cummings

Interpreting Graphs and Data

Scientists are debating what effects global climate change may have on nutrient cycles. As soil becomes warmer, especially at far northern latitudes, nutrients in the soil should become more available to plants, stimulating plant growth. One hypothesis is that more carbon will end up stored in the soil as a result, because plants will pull carbon from the atmosphere and transfer it to the soil reservoir as they shed leaves or die.* Under this hypothesis, increased flux of carbon from the atmosphere to the soil would act as negative feedback counteracting climate warming, because less carbon in the atmosphere would lead to less warming. To test whether the carbon flux actually changes in this way when nutrients are made more available in a tundra ecosystem, researchers are conducting a long-term study in Alaska.† For 20 years, they have added fertilizer to treatment plots while leaving control plots unfertilized. Recently, they estimated amounts of carbon by measuring biomass aboveground and belowground in both sets of plots. Aboveground biomass consists of living plant material, whereas belowground biomass consists mostly of nonliving organic material stored in the soil and not yet decomposed. Some of the research team's results are presented in the graph below.

*Hobbie, S. E., et al. 2002. A synthesis: The role of nutrients as constraints on carbon balances in boreal and arctic regions. *Plant and Soil* 242: 163–170. †Mack ,M. C., et al. 2004. Ecosystem carbon storage in arctic tundra reduced by long-term nitrogen fertilization. *Nature* 431: 440–443.

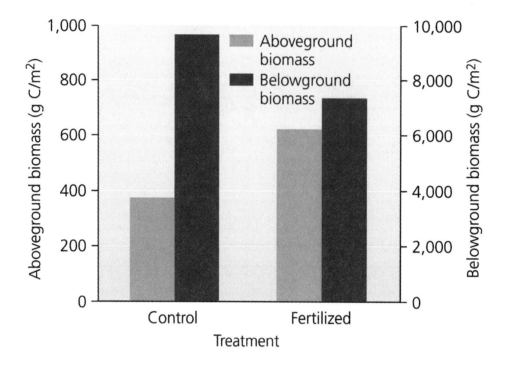

Effects of 20 years of fertilization (10 g N and 5 g P per m^2 per year) on C pools in tundra near Toolik Lake, Alaska. Differences are statistically significant for the aboveground, belowground, and total carbon pools.

1. Calculate the sizes of the aboveground, belowground, and total carbon pools (in g C/m^2) for the Control and Fertilized treatment groups.

2. What do the aboveground data indicate about the effect of fertilizer on plant growth? What do the belowground data indicate about the effect of fertilizer on organic material stored in the soil?

3. Do the data support the hypothesis that the net effect of increased nutrient availability will be to remove carbon from the atmosphere and store it in the soil? Can you suggest a different hypothesis that might explain the data better? Based on these data, would you predict that the warming of tundra soil will decrease the atmospheric concentration of CO_2 and act as negative feedback to climate change, or increase it and act as positive feedback?

Calculating Ecological Footprints

In the United States, a common dream is to own your own home, surrounded by a weed-free, green lawn. Nationwide there are about 20 million acres of lawn grass, making it the nation's largest single crop! Assuming that all of the populations indicated in the following table have lawns and will fertilize them at a typical fertilizer application rate of 45 lb of nitrogen per acre, calculate the total amount of nitrogen that will be applied to their lawns. When estimating the number of lawns, assume that the typical household includes three people.

Taxon	Percentage of total introduced	Number of species introduced
Plants	50	25,000
Mammals	0.04	
Birds	0.19	
Molluscs	0.18	
Arthropods	9	
Microbes	40	

1. Where does all of this nitrogen come from? Where does it go?

2. What other environmental impacts are caused by fertilizer production, transport, and application?

3. What steps could you take to reduce nitrogen pollution?

Case Study

Mountaintop Removal Coal Mining and GIS Technology

In the Appalachian region of the United States, a devastating mining practice is widely used to extract coal for electricity generation. Called mountaintop removal, or MTR for short, it involves using high-powered explosives to remove the tops off of entire mountains in order to harvest narrow coal seams beneath. The tops of the mountains, mainly rubble containing some coal, are left to sit, often contained by hastily constructed dams. These remaining deposits have caused large scale pollution of rivers, streams, and groundwater. They also present a risk of landslides if these precarious dams brake. You will learn more about this process in Chapter 19 of *Environment: The Science behind the Stories, 3ed.*, or Chapter 15 of *Essential Environment, 3ed.* Widely used in Kentucky, Tennessee, Ohio, Virginia, and West Virginia, this practice exemplifies how deeply tied together the ecological landscape is. As mountains change, so too do landscapes and water systems, even down to the level of water used for human consumption.

Community groups in Appalachia, as well as national environmental justice organizations, are working hard to bring this practice to an end. But in order to do so, they need accurate information about how much coal is mined in this fashion. Some of these groups employ Geographic Information Systems, or GIS, to determine how much mountaintop removal mining is happening in their communities, and where the coal derived from it is being shipped. With this information, these groups then educate the public about where electricity comes from, with the assumption that most people don't want to rely on destructively mined coal. One website that does this enables visitors to type in their zip codes and learn exactly which mountain was destroyed in order to provide electrical service. This would only be possible thanks with elaborate GIS technology.

Thanks to the help of tools like GIS, landscape ecology can be better understood. It is the hope of anti-MTR groups that with enhanced public education the practice can be brought to an end.

Resources:

Appalachian Voices: *www.appalachianvoices.org*

Search for your connection to mountaintop removal at *www.ilovemountains.org*

Activity:

Can you think of other environmental challenges that would be aided by GIS technology? Think of examples that depend on many different layers of information. Once you think of a problem, begin to sketch the different layers of the land that are impacted by the particular problem.

Questions:

1. Prior to the advent of GIS technology, how do you think ecological information was collected? Bearing in mind that many parts of the world do not have access to this technology, how might traditional methods be enhanced, or better technology made more available?

2. There is currently a loophole in federal policy that enables mountaintop removal mining to continue. Can you think of other reasons why the devastating practice continues, both at the state policy and community levels?

Topic 8: Human Population

Interpreting Graphs and Data

In 2006, the population of the United States reached 300 million. As the nation's population has increased, its demographics have shifted as well, including the numbers of people per household. This bar graph depicts the percentage of U.S. households of different sizes in 1970 and in 2000.

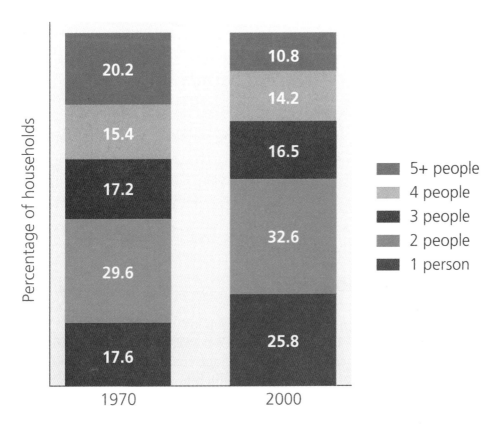

Percentage of U.S. households with one, two, three, four, and five or more people, in 1970 and in 2000. Data from U.S. Census Bureau.

1. Did household sizes generally increase, decrease, or stay the same from 1970 to 2000?

2. How many times more common were one-person households than households of five people or more in 2000?

3. What percentage of households had fewer than three people in 1970 and in 2000?

4. Describe the impacts you think these changes in household size may have had, in terms of **(a)** land use and development, and **(b)** consumption of energy and natural resources. Do you feel that household size is an important element to consider when determining the environmental impacts of population growth?

Calculating Ecological Footprints

A nation's population size and the affluence of its citizens each influence its resource consumption and environmental impact. In 2006, the world's population passed 6.5 billion. Average per capita income was $9,190 per year, and the average ecological footprint was 2.2 hectares (ha) per person. The sampling of data in the table will allow you to explore patterns in how population, affluence, and environmental impact are related.

Nation	Population (millions of people)	Affluence (per capita income, in GNI PPP)	Personal impact (per capita footprint, in ha/person)	Total impact (national footprint, in millions of ha)
Brazil	186.8	$8,230	2.1	391.9
Ethiopia	74.8	$1,000	0.8	
Japan	127.8	$31,410	4.4	
Mexico	108.3	$10,030	2.6	
Russia	142.3	$10,640	4.4	
United States	299.1	$41,950	9.6	2871.4

Data sources: Population Reference Bureau. 2006. *World population data sheet 2006*; and WWF—World Wide Fund for Nature, 2006. *Living planet report.* Gland, Switzerland: WWF.

1. Calculate the total impact (national ecological footprint) for each country.

2. Draw a graph illustrating per capita impact (on the y axis) vs. affluence (on the x axis). What do the results show? Explain why the data look the way they do.

3. Draw a graph illustrating total impact in relation to population. What do the results suggest to you?

4. Draw a graph illustrating total impact in relation to affluence. What do the results suggest to you?

5. You have just used three of the four variables in the IPAT equation, (where I is total impact, P is population, A is affluence and T is technology). Can you give one example of how the T variable might increase the total impact of the United States, and one example of how it might decrease the U.S. impact?

Case Study

Global Health Week of Action

The AIDS epidemic is having the greatest impact on human populations of any disease since the 14th-century Black Plague swept Europe and the Smallpox epidemics devastated entire civilizations of native peoples in the Americas. Today's AIDS epidemic is ravaging many parts of the developing world, especially sub-Saharan Africa, but it has claimed victims in all parts of the world. By claiming myriad victims, many of them young and highly productive members of society, HIV/AIDS disrupts demographic patterns and inhibits poor countries' abilities to develop.

Some organizations in the United States try to curb the spread of infectious disease by educating Americans about the problems. The Baltimore-based organization Americans for an Informed Democracy (AID) engages U.S. students to become educated about global health challenges, especially HIV/AIDS. It is a difficult goal, because for many students, the AIDS epidemic feels very far away. In order to overcome this, AID employs YouTube videos to get its message across. For the Global Health Week of Action from March 24-28, 2008, AID used a creative YouTube video to inspire students to participate in the week's events. Because many students already spend time on YouTube, AID calculated that more students would be exposed to their message through that medium.

Resources:

Americans for Informed Democracy: *www.aidemocracy.org*
Their video: *www.aidemocracy.org/health/weekofaction08.php*
*If you're unable to watch the video, look at this website with other ideas for students to participate in World Health Day: *www.who.int/world-health-day/toolkit/en/*

Activity:

How would you design a YouTube video that spreads the word about World Health Day? Who would be in it, how long would it be, and what would the plot elements be? Do you think creating YouTube videos is a good way to engage college students? Why or why not?

Questions:

1. With increasing knowledge in the medical community, do you think global epidemics can be prevented? Why or why not?

2. When new epidemics break out—for example, SARS in the early 2000s—how might more people be educated about the problems and solutions? Would you appreciate being better informed about such matters? Explain your answer.

3. Given what you've learned about how epidemics spread, think of what the next potential epidemic might be, and in what part of the world it might break out.

Topic 9: Soils

Causes and Consequences

CAUSES AND CONSEQUENCES

Soil erodes naturally, but human activities can greatly speed the rate of **soil erosion**, leading to diverse ecological effects and to potentially severe impacts on agricultural production. Fortunately, farmers, scientists, and policymakers have devised a number of ways to conserve soil and prevent or mitigate erosion.

Write in two causes of soil erosion in the spaces provided. Then write in two consequences (impacts on the environment, human health, or quality of life) that result from soil erosion. Finally, offer two solutions to this issue and its consequences. One cause, one consequence, and one solution have been filled in for you, providing examples.

CAUSES
- Erosion by wind
- _____
- _____

→ Soil Erosion →

CONSEQUENCES
- Loss of topsoil
- _____
- _____

↓

SOLUTIONS
- Planting shelterbelts to block wind on farmland
- _____
- _____

Sometimes solutions can have unintended consequences. Can you think of an undesired consequence that one solution to this issue might have? How might we then deal with *that* consequence?

Copyright © 2008 Pearson Education, Inc., publishing as Pearson Benjamin Cummings

Interpreting Graphs and Data

Kishor Atreya and his colleagues at Kathmandu University in Nepal conducted a field experiment to test the effects of reduced tillage versus conventional tillage on erosion and nutrient loss in the Himalayan Mountains in Central Nepal. The region in which they worked has extremely steep terrain (with an average slope of 18%), and receives over 138 cm (55 in.) of rain per year, with 90% of it falling between May and September. Atreya's team measured the amounts of soil, organic carbon, and nitrogen lost from the research plots (which were not terraced) over the course of a year. Some of their results are presented in the graph on the next page.

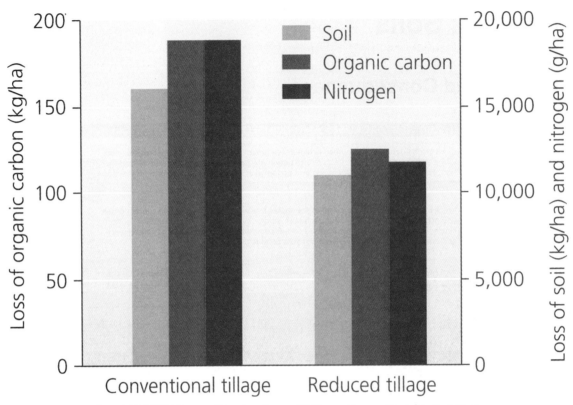

Annual soil and nutrient losses in plots under conventional and reduced tillage systems. All reduced tillage values are significantly different from their conventional tillage counterparts. Data from Atreya, K., et al. 2006. Applications of reduced tillage in hills in Central Nepal. *Soil & Tillage Research* 88:16–29.

1. Under the conditions of the study reported above, how much soil, organic carbon, and nitrogen would be saved annually in fields with reduced tillage relative to fields with conventional tillage? Express your answers both in absolute units and as percentages.

2. Given that annual crop yields in each of the study plots were approximately 4 metric tons/ha, what is the ratio of soil lost to crop yield under conventional tillage? Under reduced tillage?

54

3. Is reduced tillage a sustainable management practice for Nepalese farmers? If so, what data from the study above would you cite in support of your answer? If not, or if you cannot say, then what concerns raised by the data in the graph on the previous page would need to be addressed, or what additional data would be needed, to answer the question?

Calculating Ecological Footprints

In the United States, approximately 6 pounds of topsoil are lost for every 1 pound of grain harvested. Erosion rates vary greatly with soil type, topography, tillage method, and crop type. For simplicity let us assume that the 6:1 ratio applies to all plant crops and that a typical diet includes 1 pound of plant material or its derived products (sugar, for example) per day. In the first two columns of the table, calculate the annual topsoil losses associated with growing this food for you and for other groups, assuming the same diet.

	Plant products consumed (lb)	Soil loss at 6:1 ratio (lb)	Soil loss at 4:1 ratio (lb)	Reduced soil loss at 4:1 relative to 6:1 ratio (lb)
You	365	2,190	1,460	730
Your class				
Your state				
United States				

1. Improved soil conservation measures reduced erosion by one-third from 1982 to 1997. If additional measures were again able to reduce the current rate of soil loss by a third, the ratio of soil lost to grain harvested would fall from 6:1 to 4:1. Calculate the soil losses associated with food production at a 4:1 ratio, and record your answers in the third column of the table.

2. Calculate the amount of topsoil hypothetically saved by the additional conservation measures in question 1, and record your answers in the fourth column of the table.

3. Define a "sustainable" rate of soil loss. Describe how you might determine whether a given farm was practicing sustainable use of soil.

Case Study

Southern Illinois University
Carbondale, IL
Founded 1869
Undergraduate enrollment: 16,193

There's more to waste reduction than recycling bottles, cans, and newspaper. For example, Southern Illinois University at Carbondale (SIUC) employs two million waste reducers: red worms.

Worms help reduce the amount of waste that ends up in landfills. Andilee Warner, recycling and solid waste coordinator at SIUC, wanted to use worms for this purpose. At her school, more than 1,200 pounds of food waste was generated each day. Warner knew that by composting, these leftovers become food for worms, not destined for landfills. Composting conducted with the aid of worms is called vermicomposting. Not only do worms speed up the natural decomposition of organic matter, but the compost produced is high in nutrients and serves as a potent natural fertilizer.

The appetite of these worms defies belief. A mere one pound of worms, which amounts to approximately 800 to 1,000, can eat up to half a pound of leftover food each day. Not only does the resulting compost divert large amounts of food waste, but when added to soil, it retains moisture and increases plant yields.

It's one thing for students to become interested in macro-level sustainability efforts, but another thing entirely to focus on the soil level, digging around in the dirt and working with worms! An active soil life is rarely noted on a college campus--how would you encourage your fellow students to, on a personal level, start noticing the ground beneath their feet? Southern Illinois University may be on to something: waste reduction and soil education at the same time.

Resources:

Southern Illinois University website: *www.siuc.edu*

Activity:

Get into small groups. Within your group, imagine you are going to propose a vermicomposting program for your campus. Who would you need to talk to? What happens to food wasted on your campus? What could your campus do with the vermicompost produced? How could you use such a program to educate students, faculty, and staff on a campus-wide basis?

Questions:

1. Do you know anybody that has a composting system at their home? If you do, how does it work and what do they do with the compost? If you don't, why do you think composting is not more common?

2. Would you consider starting a vermicomposting project at your current place of residence? Why or why not? Do you think you could change your mind in the future? Explain your answer.

3. Why is so much food wasted in the United States today? As food prices continue to rise, do you think less food will be wasted? Why or why not?

Topic 10: Agriculture and the Food We Eat

Interpreting Graphs and Data

In the year 2000, over 80 million metric tons of nitrogen fertilizer were used in producing food for the world's 6 billion people. Food production, use of nitrogen fertilizers, and world population all had grown over the preceding 40 years, but at somewhat different rates. During this time, food production grew slightly faster than population while relatively little additional land was converted to agricultural use. Fertilizer use grew most rapidly.

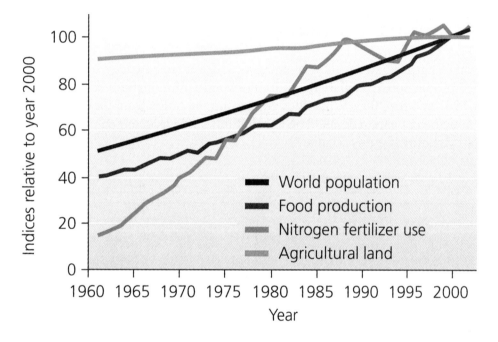

Global food production, nitrogen fertilizer use, human population, and land in agriculture, 1961–2003, relative to 2000 levels (2000 = 100). Data from U.N. Food and Agriculture Organization.

1. Express the year 2003 values of the four graphed indices as percentages of the value of each index in 1961.

2. Calculate the ratio of the food production index to the nitrogen fertilizer use index in 1961 and in 2003. What does comparing these two ratios tell you about how the efficiency of nitrogen use in agriculture has changed? Is this an example of the principle of diminishing returns?

3. As world population has grown, so has the demand for food, yet we have devoted little additional land to food production. Calculate the ratio of the agricultural land index to the population index for 1961 and for 2003. What does comparing the ratios from these two years tell you about how the per capita demand on agricultural land has changed over time? To what factors would you attribute this change?

Calculating Ecological Footprints

As food production became more industrialized during the 20th century, several trends emerged. One trend, documented in this chapter, was a loss in the number of varieties of crops grown. A second trend was the increasing amount of energy expended to store food and ship it to market. In the United States today, food travels an average of 1,400 miles from the field to your table. The price you pay for the food covers the cost of this long-distance transportation, which in 2004 was approximately one dollar per ton per mile. Assuming that the average person eats 2 pounds of food per day, calculate the food transportation costs for each category in the table on the next page.

Consumer	Daily cost	Annual cost
You	$1.40	$511
Your class		
Your town		
Your state		
United States		

1. What specific challenges to environmental sustainability are imposed by a food production and distribution system that relies on long-range transportation to bring food to market?

2. A study by Pirog and Benjamin* noted that locally produced food traveled only 50 miles or so to market, thus saving 96% of the transportation costs. Locally grown foods may be fresher and cause less environmental impact as they are brought to market, but what are the disadvantages to you as a consumer in relying on local food production? Do you think the advantages outweigh those disadvantages?

*Pirog, R., and A. Benjamin. 2003. *Checking the food odometer: Comparing food miles for local versus conventional produce sales to Iowa institutions.* Ames, IA: Leopold Center for Sustainable Agriculture, Iowa State University.

3. What has happened to gasoline prices recently? How would future increases in the price of gas affect your answers to the preceding questions?

Case Study

Farmers' Markets

Biotechnology and industrial agriculture raise many questions about food quality, food safety, and larger ecological concerns. One of the most widely recognized alternatives to conventional food systems is local, organic farms, whose produce is mainly sold at farmers' markets or through community-supported agriculture programs. Farmers' markets represent a nexus between individual needs and community needs: They satisfy individuals' interest in organic and locally grown food, and provide a location for community connections and for improving the local economic base for food production. There are social benefits as well. Examining data on social behavior at farmers' markets as compared to supermarkets, environmental author Bill McKibben concluded that farmers' market shoppers engaged in 10 times as many conversations per visit as shoppers in supermarkets.

One of the most celebrated farmers' markets in North America is located in Santa Fe, New Mexico. It began in the late 1960s, and it assures buyers that most products sold by its vendors are locally grown by the people selling them. One hundred percent of the vegetables, fruits, and nursery plants for sale at the Santa Fe farmers' market are grown in northern New Mexico. Vendors of craft items and processed foods also strive for this goal, and at least 80% of those for sale are also local.

Resources:

October 31, 2007, "Bill McKibben: By Itself, Growth Isn't Enough," US News and World Report. *www.usnews.com/articles/business/economy/2007/10/31/bill-mckibben-by-itself-growth-isnt-enough.html*

"About the Santa Fe Farmers' Market:" *www.santafefarmersmarket.com/about/*

Community Supported Agriculture: *www.localharvest.org/csa*

Activity:

Is there a farmers' market in your community? Have you ever visited? If so, discuss with a group of your classmates whether you would visit again, and why or why not. If you haven't visited a farmers' market, make a list of the reasons why. Because farmers' markets are often publicized through word of mouth, how could college students be made more aware of them?

Questions:

1. What are the main obstacles for students shopping at farmers' markets?

2. If your school purchased local foods for its dining halls, would you want them to be labeled as such? What about for organic foods? Why or why not?

3. Would you be in favor of government subsidies to promote local food production, instead of larger, industrial-scale agriculture? Explain your answer.

Topic 11: The Importance of Protecting Biodiversity

Interpreting Graphs and Data

Habitat alteration is the primary cause of present-day biodiversity loss, and of all human activities, the one that has altered the most habitat is agriculture. Between 1850 and 2000, 95% of the grasslands of the Midwestern United States were converted to agricultural use, and industrialized farming practices replaced diverse natural communities with greatly simplified ones. The vast monocultures of industrialized agriculture produce bountiful harvests, but the intensive use of water, chemical fertilizers, and pesticides bring substantial costs in lost ecosystem services. Recently, researchers surveyed the scientific literature, reviewing all research that compared the effects on biodiversity of organic farming versus conventional industrialized farming practices. The scientists predicted organic farming would benefit most organisms because of reduced chemical pollution. The graph on the following page shows the overall number of studies that showed positive effects (such as increases in abundance or species richness) and negative effects of organic farming on biodiversity.

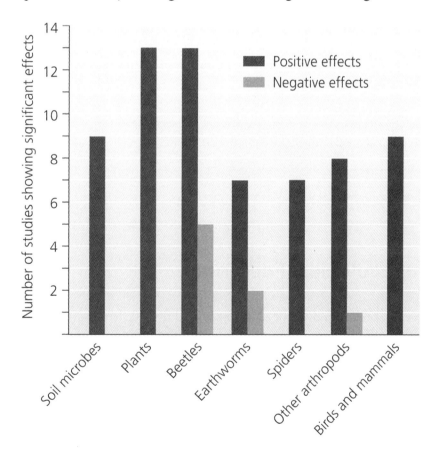

Numbers of scientific studies reporting positive and negative effects on biodiversity of organic agriculture (red bars) versus conventional farming practices (orange bars). Data from Hole, D., et al. 2005. Does organic farming benefit biodiversity? *Biological Conservation* 122: 113–130.

1. Overall, how many studies showed a positive effect of organic farming on biodiversity? How many studies reported a negative effect?

2. For which group or groups of organisms is evidence of positive effects the strongest? Reference the numbers to support your choice(s).

3. Recall the ecosystem services provided by biodiversity. What services do the group(s) you chose in Question 2 provide?

Calculating Ecological Footprints

Of the five major causes of biodiversity loss discussed in your text, habitat alteration arguably has the greatest impact. In their 1996 book introducing the ecological footprint concept, authors Mathis Wackernagel and William Rees present a consumption/land-use matrix for an average North American. Each cell in the matrix lists the number of hectares of land of that type required to provide for the different categories of a person's consumption (food, housing, transportation, consumer goods, and services). Of the 4.27 hectares required to support this average person, 0.59 hectares are forest, with most (0.40 hectares) being used to meet the housing demand. Using this information, calculate the missing values in the table on the next page.

	Hectares of forest used for housing	Total forest hectares used
You	0.40	0.59
Your class		
Your state		
United States		

Data from Wackernagel, M., and W. Rees. 1996. *Our ecological footprint: Reducing human impact on the earth.* British Columbia, Canada: New Society Publishers.

1. Approximately two-thirds of the forests' productivity is consumed for housing. To what use(s) would you speculate that most of the other third is put?

2. If the harvesting of forest products exceeds the sustainable harvest rate, what will be the likely consequence for the forest? For communities surrounding the forest?

3. What impacts would you expect on biodiversity in each of the following cases, and why?

    a. The cutting of small plots of forest within a large forest.

b. The clear-cutting of an entire forest.

c. The clear-cutting of an entire forest followed by planting of a monocultural plantation of young trees.

Does the spatial scale at which you judge the effects on biodiversity affect your answer? In what ways?

Case Study

University of Minnesota
St. Paul, Minnesota
Founded 1851
Undergraduate enrollment: 28,703

The conservation and preservation ethics have deep roots in U.S. history, with key figures including John Muir, Gifford Pinchot, and Theodore Roosevelt. But conservation biology—the study of the natural world, in order to better preserve it—is a relatively new discipline. Many elements of conservation biology involve protecting biodiversity, including studying habitats and behavior of different species, and the effects of environmental change on these species. In some cases, a particular species may face the threat of extinction, in which case conservation biology provides a framework with which to monitor and protect the species' population. Because of all these elements and more, the field is interdisciplinary, integrating both natural and social sciences.

The University of Minnesota was well positioned to help further this discipline, thanks to strong programs in ecology, social sciences, fisheries, aquatic biology, and environmental social sciences. The University of Minnesota founded its graduate program in conservation biology in 1990. Today, the program is among the most comprehensive in the country, offering M.S., Ph.D., and joint J.D. degrees, as well as a minor. Faculty and students conduct research on terrestrial and aquatic ecosystems in the United States and throughout the world.

One notable aspect of the program is that students partner with local, state, and federal conservation agencies, as well as national and international non-governmental organizations, to gain hands-on experience in the conservation and management of biodiversity. Student research projects cover many topics and are conducted around the world, from Africa to Nepal to South Dakota's Badlands. Examples include everything from the effects of walleye fishing tournaments on walleye survival, to the role of earthworms in forest degradation, to lion-human conflict in Tanzania.

Resources:

University of Minnesota Conservation Biology program: *www.consbio.umn.edu*

University of Minnesota Woodcock Preservation program: *fwcb.cfans.umn.edu/coop/projects/woodcock/index.html*

Activity:

List several different habitats that are found in your community. Are these areas being protected? If they are, in what way? If they are not, why not? If you were to become aware of a valuable natural area in need of protection, who would you talk to and how would you go about convincing others to join you?

Questions:

1. What disciplines in environmental studies overlap with conservation biology? List as many as you can think of.

2. Many people are unaware of the importance of protecting biodiversity. How could you make such people aware of the value of protecting different species and their habitats? How might your efforts differ for an urban area, versus a suburban area, versus a rural area?

3. Is protecting all species a realistic goal for conservation biologists? Explain your answer.

Topic 12: Forest Management

Causes and Consequences

CAUSES AND CONSEQUENCES

Although we are making some strides toward sustainable forestry, we are still deforesting 13 million hectares (32 million acres) globally each year. When balanced against forest regrowth, this **deforestation** gives us an annual net loss of 7.3 million hectares (18 million acres). While deforestation has a number of causes, there are steps we can take to reduce the loss of forests and the consequences that deforestation brings.

Write in two causes of deforestation in the spaces provided. Then write in two consequences (impacts on the environment, human health, or quality of life) that result from deforestation. Finally, offer two solutions to this issue and its consequences. One cause, one consequence, and one solution have been filled in for you, providing examples.

CAUSES
- Expansion of farming and grazing
- _____
- _____

Deforestation

CONSEQUENCES
- Soil erosion
- _____
- _____

SOLUTIONS
- Find substitutes for wood products
- _____
- _____

Sometimes solutions can have unintended consequences. Can you think of an undesired consequence that one solution to this issue might have? How might we then deal with *that* consequence?

Copyright © 2008 Pearson Education, Inc., publishing as Pearson Benjamin Cummings

Interpreting Graphs and Data

The invention of the movable-type printing press by Johannes Gutenberg in 1450 stimulated a demand for paper that has only increased as the world population has grown. The 20th-century invention of the xerographic printing process used in photocopiers and laser printers has accelerated our demand for paper, with most raw fiber for paper production coming from wood pulp from forest trees.

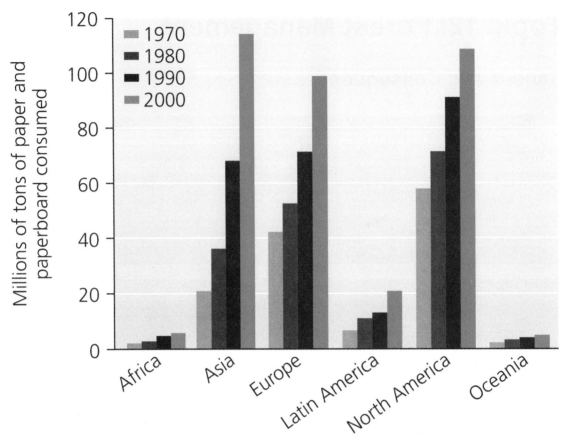

Global consumption of paper and paperboard, 1970–2000. Data from the Food and Agriculture Organization of the United Nations.

1. How many millions of tons of paper and paperboard were consumed worldwide in 1970? 1980? 1990? 2000?

2. By what percentage did worldwide consumption of paper and paperboard increase from 1970 to 1980? From 1980 to 1990? From 1990 to 2000?

3. Name three steps that your school could take to reduce its paper consumption.

Calculating Ecological Footprints

The average North American uses over 300 kg (660 lb) of paper and paperboard per year. Using the estimates of paper and paperboard consumption for each region of the world for the year 2000—as shown in the graph in the "Interpreting Graphs and Data" section—calculate the per capita consumption of paper and paperboard for each region of the world using the population data in the table.

	Population in 2000 (millions)*	Total paper consumed in 2000 (millions of tons)	Per capita paper consumed in 2000 (pounds)
Africa	840	6	14
Asia	3,766		
Europe	728		
Latin America	531		
North America	319		
Oceania	32		
World	6,216		~114

*Data source: Population Reference Bureau.

Copyright © 2008 Pearson Education, Inc., publishing as Pearson Benjamin Cummings

1. How much paper would North Americans save each year if we consumed paper at the rate of Europeans?

2. How much paper would be consumed if everyone in the world used as much paper as the average European? As the average North American?

3. Why do you think people in other regions consume less paper, per capita, than North Americans? Name three things you could do to reduce your paper consumption.

Case Study

Greenpeace Kleercut Campaign

Greenpeace is one of the largest environmental organizations in the world. Known largely for its radical tactics to protect whales, it also engages in corporate campaigning, in which particular companies' practices are highlighted in order to initiate industry-wide changes. In an effort to combat old-growth logging, Greenpeace has chosen Kimberly Clark as a major target. With the slogan: "Kleercut: Wiping Away Ancient Forests," Greenpeace aims to poke fun at the Kleenex brand of tissues, made by Kimberly Clark. The Kleercut campaign aims to educate consumers about where paper products come from and how to avoid purchasing those that harm forests.

According to Greenpeace's research, in North America, little of Kimberly-Clark's disposable paper products comes from recycled sources--only around 19%. The rest is made from pulp derived from forests, many of which are thousands of years old. In contrast, some paper companies use 100% recycled paper in their products. One example is Burlington, Vermont-based Seventh Generation, a successful business in its own right.

Another standard for non old growth forest paper is certification from the Forest Stewardship Council (FSC). FSC is well respected because it utilizes a multidisciplinary set of principles, including economic, social, and environmental issues. Its standards are now applied in more than 57 countries, and Greenpeace and other pro-forest organizations endorse the FSC criteria.

In order to promote its campaign, and reduce old growth logging, Greenpeace works with students to encourage their campuses to buy paper products that are "forest-friendly." According to the Kleercut website: "if each Canadian household replaced a box of virgin fiber facial tissue with a box of 100% recycled fiber tissue, it could save 11,654 trees, 1100 cubic yards of landfill space equal to 48 garbage trucks, and 4.2 million gallons of water." Imagine that number multiplied on the macro level at a college campus.

Resources:

Kleercut information: *www.kleercut.net*

Seventh Generation: *www.seventhgeneration.com*

Forest Stewardship Council: *www.fscus.org*

Kleenex: *www.kleenex.com/NA/FAQ.aspx#E03*

Activity:

What brands of paper products does your school use? Check out the wrapper on the tissues and toilet paper in the nearest restroom, and look near a printer for the source of printing paper. How much, if any, recycled content is included? Would you support an increase in student fees to pay for FSC certified paper, or paper with higher recycled content? Imagine that your state passed an ordinance banning the importation of old-growth logged paper products. How would that affect your campus? Discuss these questions in small groups and write your answers below.

Questions:

1. Do you think a corporate campaign such as Kleercut is an effective way to reduce old growth logging? If yes, why? If not, what do you think would be more effective?

2. How much more per package of toilet paper would you be willing to pay to buy recycled or paper certified by the FSC? Explain your answer.

3. Think of all of the different types of paper you consume in a day. How can you reduce the amount of paper that you are using? Are you willing to make these changes? Why or why not?

Topic 13: Urban Land Use

Interpreting Graphs and Data

In the graph below, urban population density is used as an indicator of sprawl (lower density = more sprawl), and carbon emissions per capita provide some measure of the environmental impact of the transportation system or preferences for each of the cities represented.

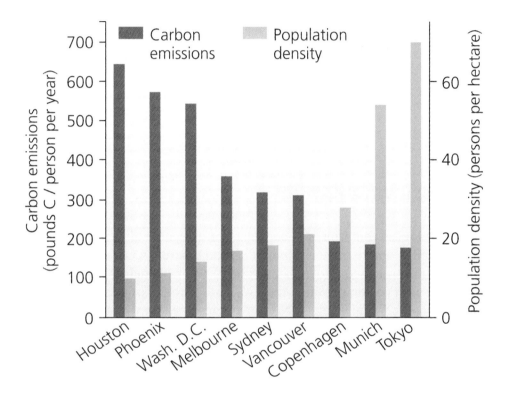

Population density versus carbon emissions from transportation in 1990. Data from Kenworthy, J., et al. 1999. *An international sourcebook of automobile dependence in cities.* Boulder, CO: University Press of Colorado, as cited by Sheehan, M. O. 2002. *What will it take to halt sprawl?* Washington DC: Worldwatch Institute.

1. Describe the relationship between urban density and carbon emissions, as shown in the graph.

2. Assuming that the standard of living is similar in these cities, to what might you attribute the relationship described in your answer to question 1?

3. If zoning ordinances slowed urban sprawl and resulted in a doubling of urban population density in a city like Houston, Texas, how would you predict that carbon emissions per capita in that city might change?

Calculating Ecological Footprints

One way of altering your ecological footprint is to consider transportation alternatives. Each gallon of gasoline is converted to approximately 20 lb of carbon dioxide (CO_2) during combustion, and this CO_2 is then released into the atmosphere. The table below lists typical amounts of CO_2 released for each person per mile, through various forms of transportation, assuming typical fuel efficiencies. For an average North American person who travels 12,000 miles per year, calculate and record in the table the CO_2 emitted yearly for each transportation option, and the reduction in CO_2 emissions that one could achieve by relying solely on each option.

	CO_2 per person per mile	CO_2 per person per year	CO_2 emission reduction	Your estimated mileage per year	Your CO_2 emissions per year
Automobile (driver only)	0.825 lb	9,900 lb	0		
Automobile (2 persons)	0.413 lb				
Automobile (4 persons)	0.206 lb				
Vanpool (8 persons)	0.103 lb				
Bus	0.261 lb				
Walking	0.082 lb				
Bicycle	0.049 lb				
				Total = 12,000	

1. What transportation option will give you the most miles traveled per unit of carbon dioxide emitted?

2. Clearly, it is unlikely that any of us will walk or bicycle 12,000 miles per year or travel only in vanpools of eight people. In the last two columns of the table, estimate what proportion of the 12,000 annual miles you think that you actually travel by each method, and then calculate the CO_2 emissions that you are responsible for generating over the course of a year. Which transportation option accounts for the most emissions for you?

3. How could you reduce your CO_2 emissions? How many pounds of emissions do you think you could realistically eliminate over the course of the next year? Would you be willing to do these things? Why or why not?

Case Study

Effective Public Transit Systems
Curitiba, Brazil

Nestled in Brazil's Parana region, with growth driven by the cattle trade, Curitiba, Brazil does not seem a likely spot to find one of the world's most advanced public transit systems. But unlikely as it may be, Curitiba is world-renowned for its bus system, and has been for three decades.

The bus system exemplifies a model Bus Rapid Transit (BRT) system. In many cities worldwide, buses are slow, erratic, crowded, and therefore not widely used. In Curitiba, city transit officials made innovations to scheduling, stations, cost, and boarding systems in order to overcome many of these persistent problems. BRT buses run frequently and keep to a reliable schedule. Stations are designed to be user-friendly. Like a subway system, fares are collected prior to boarding, and bus passengers are not impeded by car traffic—but unlike a subway, this all happens aboveground. As a result, around 70% of Curitiba commuters use the BRT, which reduces traffic congestion and pollution.

One difference between Curitiba's system and others is how passengers board the bus. Instead of standing in a single line on the ground, passengers wait on a raised platform to board. In this way, more people can board the bus more quickly, thus reducing time spent with the engine idling, and also reducing wait times—which in turn encourages more people to take the bus.

With this system, Curitiba has put itself on the map for urban sustainability, and its transit system is one of the perennial examples hailed at green urban design shows.

Resources:

Urban Habitat: *www.urbanhabitat.org/node/344*

McKibben, Bill, "Curitiba: A Global Model for Development:" *www.commondreams.org/views05/1108-33.htm*

Activity:

You may not live in Curitiba, but you can still imagine better use of public transit in your community. Design an event for your school in which students are given incentives for not driving to campus on a particular day. Alternative transportation could include bicycling, walking, riding the bus, or other means you deem appropriate. When would you host the day? What incentives would you offer? How could this event be used to inspire more lasting behavioral changes among the people involved?

Questions:

1. Do you ever take alternative transportation to school? If you do, what are the benefits? If you don't, what would need to happen for you to do so?

2. Given rising concern about climate change, as well as rising gas prices, do you think the use of alternative transportation will increase in your community? Why or why not?

3. Should federal, state, and local governments invest more money in public transportation (bus or train) and the creation of bike routes? Explain your answer.

Topic 14: Toxicology

Causes and Consequences

CAUSES AND CONSEQUENCES

In the wake of industrialization, our society has manufactured many thousands of products, materials, and chemical substances. These have helped bring us higher standards of living and more comfortable lives, but they have also introduced many new **toxic substances in the environment**. There is plenty we can do, however, to minimize concentrations of artificial toxic substances in our environment and to limit our exposure to them.

Write in two causes of toxic substances in the environment in the spaces provided. Then write in two consequences (impacts on the environment, human health, or quality of life) that result. Finally, offer two solutions to this issue and its consequences. One cause, one consequence, and one solution have been filled in for you, providing examples.

CAUSES
Commercial production of many chemicals
_____
_____

Toxic Substances in the Environment

CONSEQUENCES
Contamination of soil, air, and water
_____
_____

SOLUTIONS
Consumer choice of low-toxicity products
_____
_____

Sometimes solutions can have unintended consequences. Can you think of an undesired consequence that one solution to this issue might have? How might we then deal with *that* consequence?

Copyright © 2008 Pearson Education, Inc., publishing as Pearson Benjamin Cummings

Interpreting Graphs and Data

To minimize their exposure to ultraviolet (UV) radiation and thus reduce their risk of skin cancer, people have increased their use of sunscreen lotions in recent decades. Recently, however, some research has shown that chemicals in sunscreens may themselves pose some risk to human health. The compounds most commonly used as UV protectants are fat soluble, environmentally persistent, and prone to bioaccumulation. Moreover, they exhibit estrogenic effects in laboratory rats (see Schlumpf, et al., 2001, as cited in the source note to the graph on the next page). Although the benefits of sunscreen use are substantial, the possible risks are not yet well understood. A hypothetical trade-off between the risk factors of UV exposure and sunscreen use illustrates the balancing act known as risk management.

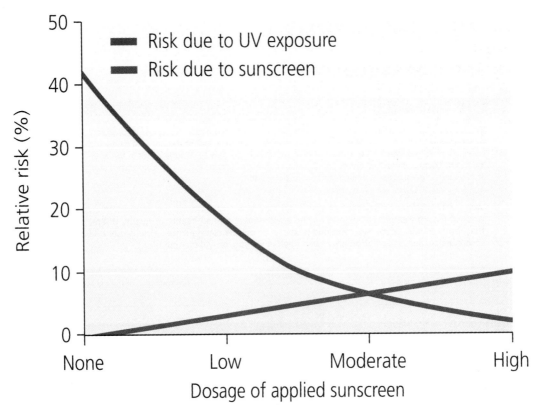

Hypothetical risk distributions for individuals using an estrogenic sunscreen to prevent skin cancer. Schlumpf, M., et al. 2001. *In vitro* and *in vivo* estrogenicity of UV screens. *Environmental Health Perspectives* 109: 239–244.

1. What dosage of applied sunscreen on the graph corresponds to the greatest risk due to UV exposure? What dosage corresponds to the greatest risk due to chemicals in the sunscreen? Which of these two points on the graph is associated with the greater risk?

2. What dosage of applied sunscreen on the graph corresponds to the least risk due to UV exposure? What dosage corresponds to the least risk due to chemicals in the sunscreen? Which of these two points is associated with the greater risk?

3. The total risk to the individual is the sum of the two individual risks. What point on the graph corresponds to the greatest total risk? What sunscreen dosage corresponds to the least total risk? Based on the data shown here, how much sunscreen would you choose to apply the next time you go to the beach? Is there any other information you'd like to know before you change the way you use sunscreen? Can you think of any other cases that illustrate this sort of trade-off between dose dependent risk factors?

Calculating Ecological Footprints

In 2001, the population of the United States was approximately 285 million, and the world's population totaled 6.16 billion. In that same year, pesticide use in the United States was approximately 1.20 billion pounds of active ingredient, and world pesticide use totaled 5.05 billion pounds of active ingredient. Pesticides include hundreds of chemicals used as insecticides, fungicides, herbicides, rodenticides, repellants, and disinfectants. They are used by farmers, governments, industries, and individuals. In the table, calculate your share of pesticide use as a U.S. citizen in 2001 and the amount used by (or on behalf of) the average citizen of the world.

	Annual pesticide use (pounds of active ingredient)
You	4.21
Your class	
Your state	
United States	
World (total)	
World (per capita)	

1. What is the ratio of your annual pesticide use to the world's per capita average?

2. Refer to the "Calculating Ecological Footprints" section in Topic 1 of this activity book, and find the ecological footprints of the average U.S. citizen and the average world citizen. Compare the ratio of pesticide usage with the ratio of the overall ecological footprints. What is the difference, and how would you account for it?

3. Does the figure for per capita pesticide use for you as a U.S. citizen seem reasonable for you personally? Why or why not? Do you find this figure alarming or of little concern? What else would you like to know to assess the risk associated with this level of pesticide use?

Case Study

California Safe Schools
Los Angeles, CA

The fight over the use of pesticides and herbicides is often conducted inside the halls of Congress, or in court, but it takes on personal and community dimensions as well. Some individuals are rising up and saying no to pesticide spraying in their communities.

According to Pesticide Watch, scientists are linking the pesticides used inside and outside schools with respiratory problems, developmental disabilities, and various forms of cancer among schoolchildren. In a survey of 46 California school districts, 87 percent still used toxic pesticides on school grounds even while safe alternatives exist.

Los Angeles parent Robina Suwol created the California Safe Schools initiative after her son Nicholas was exposed to a toxic cloud of pesticides at his grade school and had a severe asthma attack. She spent time educating her community members about the dangers of pesticide spraying at schools, and also influenced policy. Through her work with California Safe Schools, Suwol was instrumental in passing a trailblazing least-toxic Integrated Pest Management program through the Los Angeles Unified School District.

In 2007, Suwol won an award from the Environmental Protection Agency for her efforts to implement the Precautionary Principle.

Resources:

Food and Water Watch: *www.foodandwaterwatch.org*

Pesticide Watch: *www.pesticidewatch.org*

California Safe Schools: *www.calisafe.org*

Activity:

Get in small groups and discuss the following questions. Are you concerned about pesticide exposure on your campus? Are there agricultural sites on your campus or nearby that might contribute to pesticide exposure? What could be other sources of exposure near your campus? As a group, list four things your campus could do to reduce students' risk of pesticide exposure.

Questions:

1. Does your campus use pesticides to control weeds and/or insects? Why or why not? If you don't know, how could you find out?

2. Over the years, many chemicals found in pesticides have been banned from use in the United States. However, in other countries, usage continues. Why would these countries continue to use such chemicals?

3. If you were planting a garden or tending your lawn, would you use chemical pesticides? Why or why not? If not, what methods would you use?

Topic 15: Freshwater Resources

Causes and Consequences

CAUSES AND CONSEQUENCES

As the world's population and our demand for water rise, we are tapping groundwater and surface water sources, often at unsustainable rates. The resulting **fresh water depletion** is having ecological and social impacts in many areas. However, we have at our disposal many proven strategies for avoiding water shortages and making our water use more sustainable.

Write in two causes of fresh water depletion in the spaces provided. Then write in two consequences (impacts on the environment, human health, or quality of life) that result from fresh water depletion. Finally, offer two solutions to this issue and its consequences. One cause, one consequence, and one solution have been filled in for you, providing examples.

CAUSES
- Population growth
- _____
- _____

Fresh Water Depletion

CONSEQUENCES
- Water shortages
- _____
- _____

SOLUTIONS
- Deploy more efficient irrigation approaches
- _____
- _____

Sometimes solutions can have unintended consequences. Can you think of an undesired consequence that one solution to this issue might have? How might we then deal with *that* consequence?

Copyright © 2008 Pearson Education, Inc., publishing as Pearson Benjamin Cummings

Interpreting Graphs and Data

Close to 75% of the fresh water used by people is used in agriculture, and about 1 of every 14 people live where water is scarce, according to a review by hydrologist J. S. Wallace. By the year 2050, scientists project that two-thirds of the world's population will live in water-scarce areas, including most of Africa, the Middle East, India, and China. How much water is required to feed over 6 billion people a basic dietary requirement of 2,700 calories per day? The answer depends on the efficiency with which we use water in agricultural production and on the type of diet we consume.

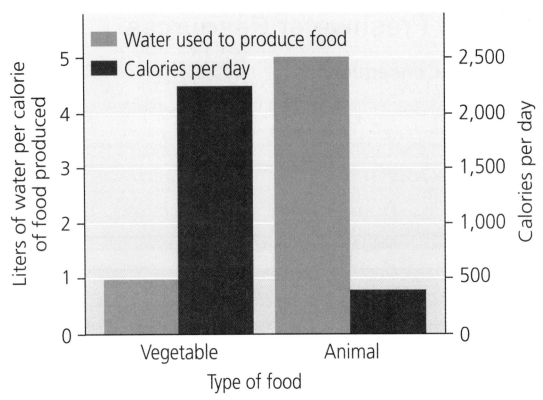

Amount of water needed to produce vegetable and animal food (orange), and global average calories per day consumed of vegetable and animal food (red). Data from Wallace, J. S. 2000. Increasing agricultural water use efficiency to meet future food production. *Agriculture, Ecosystems and Environment* 82: 105–119.

1. How many liters of water are needed to produce 2,300 calories of vegetable food? How many liters of water are needed to produce 400 calories of animal food? How many liters of water are needed daily to provide this diet? Annually?

2. How many liters of water would be saved daily, compared to the diet in the graph, if the 2,700 calories were provided entirely by vegetables? Annually?

3. Reflect on one of the quotes at the beginning of this chapter: "Water promises to be to the 21st century what oil was to the 20th century: the precious commodity that determines the wealth of nations." How do you think the demographic pressure on the water supply could affect world trade, particularly trade of agricultural products? Do you think it could affect prospects for peace and stability in and among nations? How so?

Calculating Ecological Footprints

In the United States, the EPA estimates that household water use averages 750 liters per person per day. One of the single greatest personal uses of water is for showering. Standard showerheads dispense 15 liters of water per minute, but so-called low-flow showerheads dispense only 9 liters per minute. Given an average daily shower time of 10 minutes, calculate the amounts of water used and saved over the course of a year with standard versus low-flow showerheads, and record your results in the table below.

	Annual water use with standard showerheads (liters)	Annual water use with low-flow showerheads (liters)	Annual water savings with low-flow showerheads (liters)
You	54,750	32,850	21,900
Your class			
Your state			
United States			

Data from U.S. EPA. 1995. *Cleaner water through conservation: Chapter 1—How we use water in the United States.* EPA 841-B-95-002.

1. What percentage of personal water consumption would you calculate is used for showering?

2. How much additional water would you be able to save by shortening your average shower time from 10 minutes to 8 minutes? To 5 minutes? Are you willing to do this? Why or why not?

3. Can you think of any factors that are not being considered in this scenario of water savings? Explain.

Case Study

"Flow: For Love of Water"
Produced in the U.S. in 2007
Official Selection, Sundance Film Festival

"Flow: For Love of Water," is a documentary film produced in 2007. It tells the story of an emerging global crisis surrounding lack of access to fresh water. The film casts an especially negative light on the issue of water privatization. The director, Irena Salina, highlights stories from around the world, including African plumbers who clandestinely reconnect shantytown water pipes that were shut off due to privatization, a California scientist who forces awareness of toxic public water sources, and the CEO of a large water corporation.

The film is not unbiased, and comes across as unabashedly in favor of community-driven solutions to global water crises, and opposed to large scale water privatization. However, it does present compelling evidence about how different regions are confronting lack of access to clean drinking water, and also how even areas with plenty of access to water may lack adequate information about water quality. By all counts, challenges associated with water loom large for millions of people.

The producers of the film provide opportunities for viewers to take action and help solve the water crisis. They propose an addition to the Universal Declaration of Human Rights and aim to gain enough signatures so that it will be adopted. Here is the text:

"[Recognizing] that over a billion people across the planet lack access to clean and potable water and that millions die each year as a result, it is imperative to add one more article to the [Universal Declaration of Human Rights], the Right to Water:

We, the undersigned, respectfully call upon the United Nations to add a 31st article to the Universal Declaration of Human Rights, establishing access to clean and potable water as a fundamental human right. We believe the world will be a better place when the Right To Water is acknowledged by all nations as a fundamental human right, and that this addition to the Universal Declaration of Human Rights represents a major step toward the goal of water for all."

Resources:

"Flow: For Love of Water:" *www.flowthefilm.com*

Article 31 petition: *www.article31.org*

Activity:

Get into small groups. Imagine your campus community is in a water-rich region. A large water bottling company seeks to move into the area and use your community's groundwater in the bottling plant. However, studies show that if they do this, community members will lose a large portion of the water they need for their homes, as well as for industrial operations and agriculture. How could you use the UN Declaration of Human Rights to make your case against the plant coming in? Think of places where you and your friends could circulate the Article 31 position in order to gain support for local water resources. Write your answers in the space below.

Questions:

1. Do you think an article about water belongs in the UN Declaration of Human Rights? Why or why not?

2. What kinds of water-related challenges exist in your community? How does your community respond to such challenges?

3. Can you see evidence of poor water quality in your community? If yes, what examples can you think of? If not, explain why.

Topic 16: Marine Resources

Causes and Consequences

CAUSES AND CONSEQUENCES

Our oceans and coastlines suffer many forms of pollution from human activities on land and on the water. This **marine pollution** has impacts on human health, human economies, fish and wildlife, natural systems, and ecosystem services. Fortunately, we have many solutions at hand to address the causes and mitigate the consequences of marine pollution.

Write in two causes of marine pollution in the spaces provided. Then write in two consequences (impacts on the environment, human health, or quality of life) that result from marine pollution. Finally, offer two solutions to this issue and its consequences. One cause, one consequence, and one solution have been filled in for you, providing examples.

CAUSES
Plastic debris, discarded nets, other trash
_____
_____

Marine Pollution

CONSEQUENCES
Animals become entangled and die
_____
_____

SOLUTIONS
Prevent dumping and littering; pick up trash from beaches
_____
_____

Sometimes solutions can have unintended consequences. Can you think of an undesired consequence that one solution to this issue might have? How might we then deal with *that* consequence?

Copyright © 2008 Pearson Education, Inc., publishing as Pearson Benjamin Cummings

Interpreting Graphs and Data

The following graph presents trends in the status of North Atlantic swordfish, a highly migratory species managed directly by the National Marine Fisheries Service. The solid red line shows the mortality rate from fishing. The solid blue line indicates the biomass of the stock. The dotted lines of corresponding colors indicate the reference levels used to determine whether the stock is overfished or recovered. The graph also indicates the date when an international recovery plan was implemented.

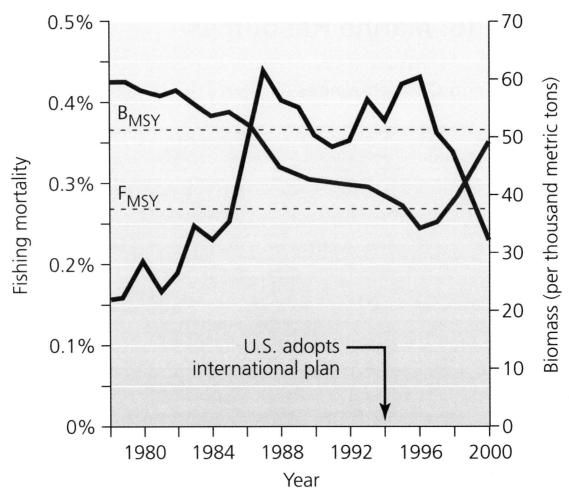

Trends in fishing mortality and stock biomass for North Atlantic swordfish, 1978–2000. Adapted from Rosenberg, A., et al. 2006. Rebuilding U.S. fisheries: Progress and problems. *Frontiers in Ecology and the Environment* 4(6), August 2006.

1. Describe the trends in swordfish stocks (1) before the United States adopted an international management plan and (2) since the plan was adopted. Describe the interactions between fishing mortality and biomass as illustrated by the graph.

2. Based on the data in the graph, predict the likely trend in swordfish production over the next 10 years, assuming no change to the status quo.

3. This graph illustrates an effort that is succeeding, but not all rebuilding plans lead to stock recovery. Beyond the existence of a plan, what actions might play a role in supporting stock recovery efforts?

Calculating Ecological Footprints

The relationship between the ecological goods and services used by individuals and the amount of *land* area needed to provide those goods and services is relatively well developed. People also use goods and services from Earth's oceans, where the concept of *area* is less useful. It is clear, however, that our removal of fish from the oceans has an impact, or an ecological footprint. The table below shows data on the mean annual per capita consumption from ocean fisheries for North America, China, and the world as a whole. Using the data provided, calculate the amount of fish each consumer group would consume per year, given the annual per capita consumption rates, for each of these three regions. Record your results in the table.

	Annual Consumption		
Consumer group	North America (21.6 kg per capita)	China (27.7 kg per capita)	World (16.2 kg per capita)
You			
Your class			
Your state			
United States	6.48×10^9 kg	8.31×10^9 kg	4.86×10^9 kg
World			

Data from U.N. Food and Agriculture Organization (FAO), Fisheries Department. 2004. *The state of world fisheries and aquaculture: 2004.* Data are for 2002, the most recent year for which comparative data are available.

1. Calculate the ratio of North America's per capita fish consumption rate to that of the world. Compare this ratio to the ratio of the per capita ecological footprints for the United States, Canada, and Mexico (see Figure 1.12, p.17 in *Environment: The Science behind the Stories*, 3 ed. or Figure 1.16, p.19 in *Essential Environment*, 3 ed.) versus the world average footprint of 2.2 ha/person/year. Can you account for similarities and differences between these ratios?

2. The population of China has grown at an annual rate of 1.1% since 1987, while over the same period fish consumption in China has grown at an annual rate of 8.9%. Speculate on the reasons behind China's rapidly increasing consumption of fish.

3. What ecological concerns do the combined trends of human population growth and increasing per capita fish consumption raise for you? What role might you play in contributing to these concerns or to their solutions?

Case Study

Monterey Bay Aquarium
Monterey, CA
Established 1984
Average annual visitors: 1.8 million

On April 13, 2008, the California Department of Fish and Game approved legislation to protect marine areas along the California Central Coast. Now, the state's marine protected areas, 29 in total, cover over 200 miles of coastline. Many interest groups participated in the policymaking process, including the Monterey Bay Aquarium Ocean Action Team. The Monterey Bay Aquarium is one of the premier aquariums in the U.S. It exhibits approximately 550 species, and is well known for its sea otters. Visitors to the aquarium, if moved by what they learn about, often opt to take action to preserve oceans. One choice for such people is to sign up to participate in the Ocean Action Team and even send a postcard to California legislators. These postcards helped to get the marine preservation legislation passed.

Resources:

Central Coast Marine Protected Areas: *www.dfg.ca.gov/mlpa/pdfs/ccmpas_brochure.pdf*

Monterey Bay Aquarium: *www.mbayaq.org*

Activity:

Imagine you are a volunteer for the Ocean Action Team at a local aquarium. How would you design a program to encourage visitors to the aquarium to get involved in your conservation work? On a separate sheet of paper, design a simple brochure to hand out to visitors that underscores why marine conservation is important.

Questions:

1. Do you think that aquariums have a responsibility to promote marine conservation? Why or why not?

2. If you have visited an aquarium, did it prompt you to spend more time thinking about conserving the oceans? If so, should aquarium visits be part of school curricula in an attempt to better educate people to the plight of marine resources? If you haven't visited an aquarium, what would prompt you to do so?

3. If most aquariums incorporated advocacy work into their programs, would it likely increase or decrease the amount and diversity of visitors? Explain your answer.

Topic 17: Air Pollution and the Atmosphere

Causes and Consequences

CAUSES AND CONSEQUENCES

Indoor air pollution receives less attention than outdoor air pollution, yet it exerts severe health impacts on millions of people. Poverty worsens indoor air pollution in the developing world; high-technology products and chemicals contribute to it in the developed world; and certain lifestyle choices can affect all of us. Fortunately, many solutions are within reach to address the causes and lessen the consequences of this environmental health risk.

Write two causes of indoor air pollution in the spaces provided. Then write two consequences (impacts on the environment, human health, or quality of life) that result from indoor air pollution. Finally, offer two solutions to this issue and its consequences. One cause, one consequence, and one solution have been filled in for you as examples.

Causes
Radon seepage into homes

Indoor Air Pollution

Consequences
Lung cancer; 20,000 deaths per year in the United States

Solutions
Measure radon with a test kit; increase ventilation

Sometimes solutions can have unintended consequences. Can you think of an undesired consequence that one solution to this issue might have? How might we then deal with *that* consequence?

Copyright © 2008 Pearson Education, Inc., publishing as Pearson Benjamin Cummings

Interpreting Graphs and Data

Since the Clean Air Act of 1970, total emissions of carbon monoxide, sulfur dioxide, nitrogen oxides, volatile organic compounds, lead, and particulate matter have dropped by over 50% while U.S. population, energy consumption, and economic productivity have all increased. As you learned from the "Interpreting Graphs and Data" feature in Topic 3 in this activity book, lead emissions resulted mostly from the combustion of leaded gasoline and dropped precipitously once leaded gasoline was phased out. Consider the other pollutants listed above as you interpret the data in the graph on the following page.

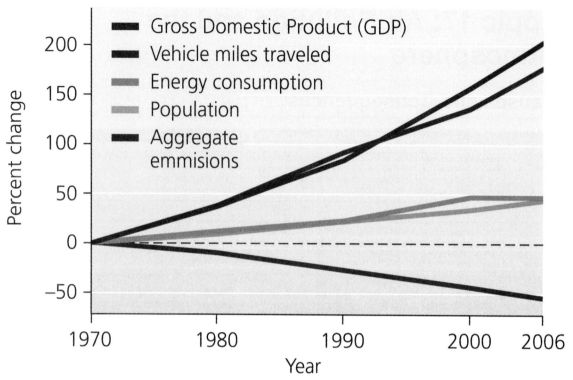

Trends from 1970 to 2006 in economic production, vehicle miles, energy consumption, population, and aggregate emissions of the six principal air pollutants monitored by the EPA. Data from U.S. EPA.

1. What are the percentage changes from 1970 to 2006 for each of the five variables graphed? Now calculate the percentage changes in the per capita values of each variable over this time period.

2. Fossil fuel combustion accounts for most of our energy consumption and is the major source of most of the pollutants named above. What is the percentage change in aggregate emissions of these six principal pollutants per unit of energy consumed in 2006, compared to 1970? What do you think accounts for this change?

3. Do you think that additional reductions in emissions are likely to result primarily from changing technology (e.g., hybrid cars, advanced catalytic converters) or from changing behavior (e.g., driving fewer miles per person per year)? Use the data in the graph to support your position.

Calculating Ecological Footprints

According to EPA data, emissions of nitrogen oxides in the United States in 2002 totaled 21,102,000 tons. Nitrogen oxides come from fuel combustion in motor vehicles, power plants, and other industrial, commercial, and residential sources, but fully 11,452,000 tons of the 2002 total came from the transportation sector. Of this amount, 7,365,000 tons came from on-road vehicles, with 3,583,000 tons of this total coming from light-duty cars and trucks. The U.S. Census Bureau estimated the nation's population to be 288,368,698 at mid-year in 2002 and projects that it will reach 308,936,000 by mid-2010. Considering these data, calculate the missing values in the table below (1 ton = 2,000 lb).

	Total NO_x emissions (lb)	NO_x emissions due to light-duty vehicles (lb)
You		
Your class		
Your state		
United States		

Data from U.S. EPA.

1. By what percentage is the U.S. population projected to increase between 2002 and 2010? Do you think that NO_x emissions will increase, decrease, or remain the same over that period of time? Why?

2. Assuming you are an average American driver, how many pounds of NO_x emissions would you prevent if you were to reduce by half the vehicle miles you travel? What percentage of your total NO_x emissions would that be?

3. How might you reduce your vehicle miles traveled by 50%? What other steps could you take to reduce the NO_x emissions for which you are responsible? Are you willing to do these things? Why or why not?

Case Study

American Lung Association
Indoor Air Pollution

One U.S. organization focused on the prevention of indoor air pollution is the American Lung Association®. According to its website, the organization is the oldest voluntary health organization in the U.S., founded in 1904 to fight tuberculosis. Now, its mission has expanded to include fighting lung disease in all its forms, emphasizing asthma, tobacco use, and environmental health.

Many of its efforts focus on outdoor air quality as it relates to asthma, but indoor air pollution also contributes to the problem. The American Lung Association® also provides resources to educate the public about reducing indoor air pollution. Some of these indoor pollutant hazards are so ubiquitous they go unnoticed. For example, dust mites in mattresses can contribute to some people's asthma symptoms, yet how does one know if a particular mattress has dust mites?
To help people with asthma cope with these challenges, the American Lung Association® provides tips for reducing indoor air pollutants during bedtime, when many symptoms occur:

1. Keep humidity below 50 percent. In the summer, that can mean using more energy in order to run an air conditioner or a dehumidifier.

2. Wash bedding every week in hot water.

3. Some types of mattresses and bedding prevent dust mite penetration. Consider purchasing those types of products, and also seal mattresses (including taping over zippers).

4. Remove wall-to-wall carpeting in favor of area rugs that can be easily removed and cleaned.

5. Wash window shades and curtains frequently.

Resources:

American Lung Association®: *www.lungusa.org/site/pp.asp?c=dvLUK9O0E&b=3474147*

Activity:

In small groups, make a list with your classmates of potential sources of indoor air pollution in your classroom and where you live. Is one list longer than the other? What are key differences, and why? Write your answers in the space below.

Questions:

1. Are you concerned about exposure to indoor or outdoor air pollutants? Explain your answer. If you are concerned, what can you do to reduce your exposure?

2. Rising concern about global climate change has focused more attention on air quality. Do you think this will result in more public concern about pollution other than that which causes climate change? Why or why not?

3. Air quality in the United States has, in some regards, improved since the passing of the Clean Air Act and its subsequent amendments. Do you think legislation regarding air pollution will be strengthened in the future? Explain your answer.

Topic 18: Global Climate Change

Interpreting Graphs and Data

We burn fossil fuels to generate electricity, to power vehicles for transportation, and as primary energy sources (nonelectricity uses, mostly for heating) in homes, businesses, and industry. For each of these uses, the accompanying graph shows trends in the emission of carbon dioxide from fossil fuel combustion in the United States.

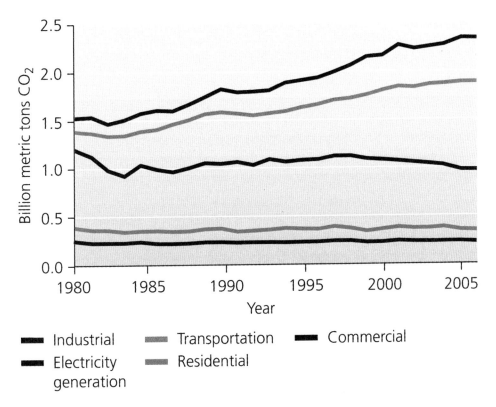

Emissions of CO_2 from fossil fuel combustion by end-use sector in the United States, 1980–2006. Data from U.S. Department of Energy, Energy Information Administration. 2007. *Annual energy review 2006.*

1. Calculate the approximate percentage changes in CO_2 emissions from transportation; electricity generation; and residential, commercial, and industrial primary energy use (mostly heating) between 1980 and 2006.

2. Between 1980 and 2006, U.S. population increased by 32% and the inflation-adjusted U.S. gross domestic product (GDP) more than doubled. What quantitative conclusions can you draw from these data about CO_2 emissions per capita? About CO_2 emissions per unit of total economic activity? Create a graph and sketch a trend line of CO_2 emissions per capita from 1980 to 2006. Now sketch a trend line of CO_2 emissions per unit of total economic activity from 1980 to 2006.

3. Imagine you are put in charge of designing a strategy to reduce U.S. emissions of CO_2 from fossil fuel combustion. Based on the data presented here, what approaches would you recommend, and how would you prioritize these? Explain your answers.

Calculating Ecological Footprints

Global climate change is something to which we all contribute, because fossil fuel combustion plays such a large role in supporting the lifestyles we lead. Conversely, as individuals, each one of us can contribute to mitigating global climate change through personal decisions and actions that affect the way we live our lives. Several online calculators enable you to calculate your own personal *carbon footprint*, the amount of carbon emissions for which you are responsible. Go to one of these, at *www.carbonfootprint.com*, follow the link for the U.S. version, take the quiz, and enter the relevant data in the table on the following page.

	Carbon footprint (kg per person per year)
World average	
Average for industrialized nations	
U.S. average	
Your footprint	
Average needed to halt climate change	
Your footprint with three changes	

1. How does your personal carbon footprint compare to that of the average U.S. resident? How does it compare to that of the average person in the world? Why do you think your footprint differs from these in the ways it does?

2. Think of three changes you could make in your lifestyle that would lower your carbon footprint. Now take the footprint quiz again, incorporating these three changes. Enter your resulting footprint in the table. By how much did you reduce your yearly emissions? How likely are you to make these changes? Why?

3. Now take the quiz again, trying to make enough changes to reduce your footprint to the level at which we could halt climate change. Do you think you could achieve such a footprint? What do you think would be an admirable yet realistic goal for you to set as a target value for your own footprint? Would you choose to purchase carbon offsets to help reduce your impact? Why or why not?

Case Study

Cornell University
Ithaca, New York
Founded 1865
Undergraduate enrollment: 13,500

Cornell University students took up the challenge of climate change at their campus, and in 2001 created an initiative called KyotoNow! This initiative encouraged Cornell to sign on to the Kyoto Protocol's carbon reduction commitment. The Kyoto Protocol requires developed nations to reduce greenhouse gas emissions to about 5% below 1990 levels by the years 2008-2012. The United States is the only advanced industrialized country not committed to these reductions, but some U.S. states and cities are nonetheless aiming to achieve Kyoto Protocol-level reductions in greenhouse gas emissions.

Why focus on the Kyoto Protocol when the U.S. government hasn't ratified it? Cornell's students saw an opportunity to tie their work on campus to something much larger in scope. At the time of the Kyoto Protocol's inception, its standards were considered a positive step forward. While today some regional agreements, such as the Regional Greenhouse Gas Initiative, call for bolder emissions reductions, the Kyoto Protocol remains unique as an international treaty. Currently, international negotiators are framing a new agreement to replace it, set to be completed in Copenhagen in December 2009.

Cornell students participate in KyotoNOW! by promoting renewable energy and energy efficiency measures on their campus. In the larger college community of Ithaca, New York, the students work to educate the public about climate change, its impacts, and how individuals can make a difference. Cornell's students are also part of a larger nationwide student network, which enables them to collaborate with partners at other schools as they work together to raise awareness about climate change and inspire people to push for positive change. These students participate in state and regional student summits, as well as national events such as Focus the Nation, a national teach-in on climate change, and Step It Up, a day of action in 2007 calling on Congress to cut carbon emissions 80% by 2050.

Resources:

KyotoNOW!: *www.rso.cornell.edu/kyotonow/index.html*

Focus the Nation: *www.focusthenation.org*

Step It Up: *www.stepitup2007.org*

Activity:

Imagine you are trying to encourage your campus to pursue the goals of the Kyoto Protocol, and that you need help. You and your fellow students decide to meet with students at Cornell to learn from them about how they put KyotoNOW! into action on a daily basis. Plan a trip to Cornell with the goal of returning to your campus with a plan of action. Who would you meet with there? What ten questions would you ask them? Would you need to see particular parts of campus in order to conduct your research (for example, witnessing exactly how renewable energy is used on campus)? How long do you think you would you need to visit?

Questions:

1. Do you think pursuing Kyoto Protocol goals on a college or university campus is realistic, given that the United States has not ratified the treaty? Why or why not?

2. By mid-2008, over 800 U.S. cities had signed on to the U.S. Mayors Climate Protection Agreement. Under this agreement, committed cities will lower their greenhouse gas emissions to 7% below 1990 levels by 2012. Do you think such an agreement will make it more likely that federal legislation to reduce greenhouse gas emissions will pass in Congress? Explain your answer.

3. In general, do you think students are effective advocates in urging college administrators to pursue far-reaching college policy, such as how energy is used? Why do you think this? How could students increase their influence over policy at your own college or university?

Topic 19: Fossil Fuels

Interpreting Graphs and Data

The fossil fuels that we burn today were formed long ago from buried organic matter. However, only a small fraction of the original organic carbon remains in the coal, oil, or natural gas that is formed. Thus, it requires approximately 90 metric tons of ancient organic matter—so-called paleoproduction—to result in just 3.8 L (1 gal) of gasoline. The graph below presents estimates of the amount of paleoproduction required to produce the fossil fuels humans have used each year over the past 250 years.

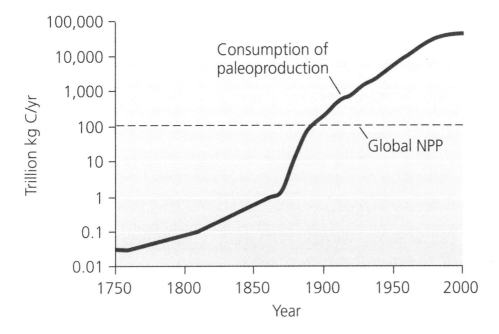

Annual human consumption of paleoproduction by fossil fuel combustion (red line), 1750–2000. The dashed line indicates current annual net primary production for the entire planet. Data from Dukes, J. 2003. Burning buried sunshine: Human consumption of ancient solar energy. *Climatic Change* 61: 31–44.

1. Estimate in what year the annual consumption of paleoproduction, represented by our combustion of fossil fuels, surpassed Earth's current annual net primary production (NPP).

2. In 2000, approximately how many times greater than global net primary production was our consumption of paleoproduction?

3. If on average it takes 7,000 units of paleoproduction to produce 1 unit of fossil fuel, estimate the total carbon content of the fossil fuel consumed in 2000. How does this amount compare to global NPP?

Calculating Ecological Footprints

Wackernagel and Rees calculated the energy component of our ecological footprint by estimating the amount of ecologically productive land required to absorb the carbon released from fossil fuel combustion. For the average American, this translates into 3 ha of his or her total ecological footprint. Another way to think about our footprint, however, is to estimate how much land would be needed to grow biomass with an energy content equal to that of the fossil fuel we burn. Assume that you are an average American who burns about 300 gigajoules of fossil fuels per year and that average terrestrial net primary productivity can be expressed as 160 megajoules/ha/year. Calculate how many hectares of land it would take to supply our fuel use by present-day photosynthetic production. A gigajoule is 10^9 (1 billion) joules; a megajoule is 10^6 (1 million) joules.

	Hectares of land for fuel production
You	1,794
Your class	
Your state	
United States	

Data from Wackernagel, M., and W. Rees. 1996. *Our ecological footprint: Reducing human impact on the Earth*. Gabriola Island, British Columbia: New Society Publishers.

1. Compare the energy component of your ecological footprint calculated in this way with the 3 ha calculated using the method of Wackernagel and Rees. Explain why results from the two methods may differ.

2. Earth's total land area is approximately 1.5×10^{10} (15 billion) ha. Compare this to the hectares of land for fuel production from the table.

3. How large a human population could Earth support at the level of consumption of the average American, if all of Earth's land were devoted to fuel production? Do you consider this realistic? Provide two reasons why or why not.

Case Study

Middlebury College
Middlebury, Vermont
Founded 1800
Undergraduate enrollment: 2,350

In January 2007, the Middlebury College Board of Trustees committed to a goal of carbon neutrality by the year 2016. The decision followed years' worth of progress on the Vermont campus, including replacing fossil fuel energy with renewable energy, encouraging students and faculty members to use less energy, and constructing energy-efficient buildings.

Remarkably, a group of 12 students proved to be the most decisive factor in the Board's decision. Calling themselves MiddShift (as in "shift into [carbon] neutral at Middlebury"), this group spent one month preparing a report and presentation to the Board. The report not only made an argument for carbon neutrality, but also laid out a plan for implementation. In addition, the students presented the findings of a detailed "energy audit" of the campus. The audit, undertaken by a student, described all campus energy use, broken down into usage categories such as heating/cooling, electricity, transportation, and waste.

The audit proved vital to the students' case, for it showed exactly how much fossil-fuel based energy the college used—thereby providing a road map for how much needed to be conserved, made more efficient, or offset with renewable energy.

Resources:

Middlebury College: *www.middlebury.edu*

MiddShift: *https://segue.middlebury.edu/index.php?action=site&site=midd_shift*

Activity: Mini Energy Audit

In small groups, look around your classroom and make a list of all the sources of energy, including electrical outlets and heating vents. Then, assess how many of these are in use, and if you can, determine the wattage of all appliances plugged into the outlets. Next, try to determine the output of the heating or cooling system in your classroom. Then, estimate how many hours each day this particular classroom is operating at peak energy usage. After you've compiled your list, multiply your classroom's usage by an estimate of the number of classrooms on your campus.

Now that you've made a rough "energy audit" of your campus, think about how your particular classroom could conserve energy, or use energy more efficiently. Are the overhead lights on a timer? Are there any appliances currently plugged in and/or turned on, but not in use? Develop a list of three simple ways your classroom could use less of your campus' share of energy.

Share your list with other groups. If time permits, develop a master list that incorporates all relevant energy-saving suggestions, and write it up on the classroom blackboard so other students may see it.

Questions:

1. The preceding case study and activity focuses on the campus level. How might you perform an energy audit for your dwelling? Would this prompt you and your roommates to conserve energy? Why or why not?

2. When city and state governments undertake plans to monitor energy use in their jurisdictions, how might they go about this? Is it absolutely necessary to perform an energy audit prior to implementing an energy-saving plan? How much information about energy use do policymakers need to know prior to crafting policy?

3. How might people be convinced that conserving energy is important? List at least five simple things people can do to start conserving energy today.

Topic 20: Conventional Energy Alternatives

Interpreting Graphs and Data

Recently, national security issues have received increased attention in energy policy discussions. Although nuclear reactors are potential targets for terrorism, they do help diversify our energy portfolio and lessen our dependence on petroleum imports. The United States produces reactor fuel but also relies on imports of uranium oxide from abroad, particularly from Canada, Australia, Russia, Kazakhstan, Uzbekistan, South Africa, and Namibia.

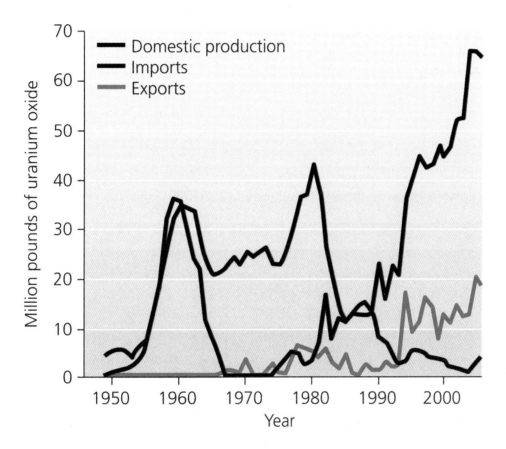

Production and trade of uranium oxide by the United States, 1949–2006. Data from Energy Information Administration. 2007. *Annual energy review 2006*. U.S. Department of Energy.

1. How much uranium oxide did the United States produce in 2005? How much did it export that year? How much did it import?

2. What was the net amount of uranium imported in 2005? How does this amount compare to domestic production for that year? How do these amounts compare to the amounts in 1980?

3. What national security concerns do these data suggest to you? How might such concerns be alleviated?

Calculating Ecological Footprints

Each of the conventional energy alternatives releases considerably less net carbon dioxide (CO_2) to the atmosphere than do any of the fossil fuels. For each fuel source in the table, calculate the net greenhouse gas emissions it produces while providing electricity to a typical U.S. household that uses 30 kilowatt-hours per day. Use data on emissions rates from each of the energy sources as provided in the figure on the following page. For each energy source, use the average of the maximum and minimum values given.

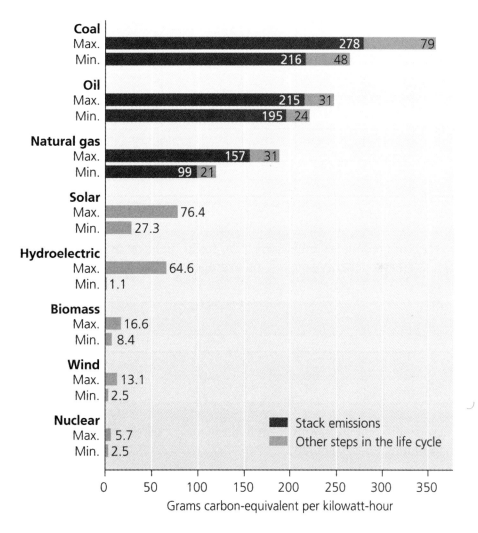

Coal, oil, and natural gas emit far more greenhouse gases than do renewable energy sources and nuclear energy. Red portions of bars represent stack emissions, and orange portions show emissions from other steps in the life cycle. Maximum and minimum values are given for each energy source. Data from Spadaro, J. V., et al. 2000. Greenhouse gas emissions of electricity generation chains: Assessing the difference. *IAEA Bulletin* 42(2).

Energy source	Greenhouse gas emission rate (g C$_{eq}$/ kW-hr)	Time period		
		1 day	1 year	30 years
Coal	311	9,330	3,405,450	102,163,500
Oil				
Natural gas				
Photovoltaic solar				
Hydroelectric				
Biomass				
Wind				
Nuclear				

1. What is the ratio of greenhouse emissions from fossil fuels, on average, to greenhouse emissions from alternative energy sources, on average?

2. Why are there significant emissions from hydroelectric power, which is commonly touted as being a nonpolluting energy source?

3. Which energy source has the lowest emission rate? Would you advocate that your nation further develop that source? Why or why not?

Case Study

Dartmouth College
Hanover, New Hampshire
"The Big Green Bus"

During the summer of 2005, a group of students at Dartmouth College took to the roads of the United States in "The Big Green Bus." Their mission involved stopping at various locations throughout the country and spreading the word about alternatives to using fossil fuel in vehicles. Their focus was on biofuels, fuels derived from plant matter and whose emissions can in some cases be considered carbon-neutral.

Today, the Big Green Bus still tours each summer, with a different group of 12 Dartmouth College students as its crew. With only 12 spots, it has become a competitive program. The students spread their message by visiting music festivals, campuses, corporate offices, summer camps, and the offices of politicians.

Their bus is a 1997 International brand diesel school bus, and they transformed both the inside and outside of the bus, not to mention how it runs. Its retrofit included removing the original bus bench seats and erecting bunk beds in their place, installing diner-style tables, and giving the exterior an eye-catching paint job.

As good-looking as the bus is, its fuel is the main selling point. The bus burns neither petroleum diesel fuel nor soy-derived biodiesel, but instead, waste vegetable oil (WVO), the cooking oil typically considered a leftover from restaurant deep-fryers nationwide. In order to burn waste vegetable oil, the Big Green Bus's engine required a few small modifications, including installing a heater to warm up the vegetable oil prior to ignition, an extra fuel tank for WVO, additional fuel lines, and a few minor additional modifications.

By burning WVO, the Big Green Bus not only avoids burning fossil fuels, but also "recycles" a common waste product. Each of their three past tours garnered extensive media coverage, putting Dartmouth College on the map for campus sustainability. The students on the bus also educated thousands of Americans about alternatives to fossil fuels.

Resources:

The Big Green Bus: *www.thebiggreenbus.org/the_bus/index.html*

Dartmouth College: *www.dartmouth.edu*

Activity:

Efforts like the Big Green Bus are one unique way to raise awareness about alternative energy sources. Get into small groups and compile a list of other ways to increase public awareness about energy issues and alternative energy sources. Do you think people today are more aware of energy issues than people were five or 10 years ago? Explain your answer.

Questions:

1. Do you think the Big Green Bus would be as interesting to people if it used biodiesel instead of waste vegetable oil? Why or why not?

2. There are disadvantages associated with biofuels, many of them involving the need to plant corn and soy for fuel instead of food. Given what you've learned of these downsides in your textbook, do you think outreach projects like the Big Green Bus are still beneficial? Why or why not?

3. What do you think about the future of transportation in the United States and the rest of the world? Do you foresee a switch to different forms of energy for transportation, or will we continue to use oil and similar products, like ethanol and biodiesel? Explain your answer.

Topic 21: New Renewable Energy Alternatives

Causes and Consequences

CAUSES AND CONSEQUENCES

Despite the appeal of renewable energy sources, they currently comprise less than 14% of our global energy supply and generate less than 19% of our electricity. Moreover, the "new renewables" account for only a very small fraction of these numbers. The fact that **renewables provide little of our energy** is a dilemma with multiple causes and consequences. However, most renewable sources are now expanding quickly, and there are many ways in which we can encourage their development and spread.

Write in two causes of the fact that renewables provide little of our energy in the spaces provided. Then write in two consequences (impacts on the environment, human health, or quality of life) that result. Finally, offer two solutions to this issue and its consequences. One cause, one consequence, and one solution have been filled in for you, providing examples.

CAUSES
Fossil fuels are rich in energy

Renewables provide little of our energy

CONSEQUENCES
Greenhouse emissions spur climate change

SOLUTIONS
Increase research and development funding for renewables

Sometimes solutions can have unintended consequences. Can you think of an undesired consequence that one solution to this issue might have? How might we then deal with *that* consequence?

Copyright © 2008 Pearson Education, Inc., publishing as Pearson Benjamin Cummings

Interpreting Graphs and Data

Of the new renewable energy alternatives discussed in your text, photovoltaic conversion of solar energy is the one that most areas of the United States could most easily adopt. The influx of solar radiation varies with time of day, time of year, and location, so all areas are not equally well suited. Today's photovoltaic technology is approximately 10% efficient at converting the energy of sunlight into electricity, but new technologies under development may increase that efficiency to as much as 40%.

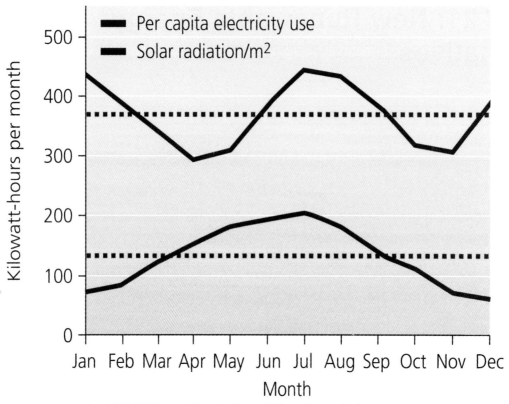

Average per capita residential use of electricity in the United States in 2004 (red line) and average influx of solar radiation per square meter for Topeka, Kansas (blue line). The dashed lines represent the yearly average values for each. Data from Renewable Resource Data Center, National Renewable Energy Laboratory, U.S. Department of Energy (DOE); and Energy Information Administration. 2005. *Annual energy review 2004.* DOE.

1. Given a 10% efficiency for photovoltaic conversion of solar energy, approximately how many square meters of photovoltaic cells would be needed to supply one person's residential electrical needs for a year, based on the yearly average values? How many square meters would be needed if efficiency were improved to 40%?

2. Given the same 10% conversion efficiency, approximately how many square meters of photovoltaic cells would be required to supply one person's residential electrical needs during the month of April? During July? How many square meters would be required to supply the average U.S. household of four people for each of those months?

3. Commercially available photovoltaic systems of this capacity cost approximately $20,000. The average cost of electricity in the United States is approximately 9¢ per kilowatt-hour. At these prices, how long would it take for the PV system to generate $20,000 worth of electricity? Calculate a combination of PV system cost and electricity cost at which the system would pay for itself in 10 years.

Calculating Ecological Footprints

Assume that average per capita residential consumption of electricity is 12 kilowatt-hours per day, that photovoltaic cells have an electrical output of 10% incident solar radiation, and that PV cells cost $800 per square meter. Now refer to the figure below, and estimate the area and cost of the PV cells needed to provide all of the residential electricity used by each group in the table on the following page.

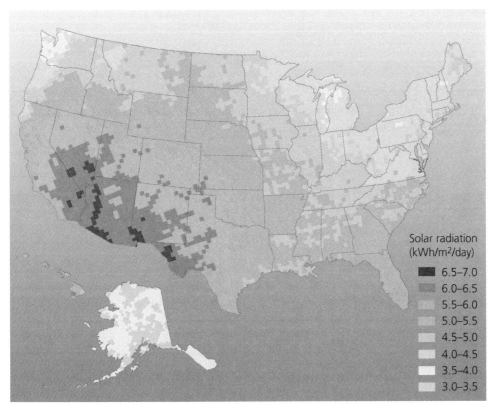

	Area of photovoltaic cells	Cost of photovoltaic cells
You	25	$20,000
Your class		
Your state		
United States		

1. What additional information do you need to increase the accuracy of your estimates for the areas in the table above?

2. Considering the distribution of solar radiation in the United States, where do you think it will be most feasible to greatly increase the percentage of electricity generated from photovoltaic solar cells?

3. The purchase price of a photovoltaic system is considerable. What other costs and benefits should you consider, in addition to the purchase price, when contemplating "going solar"?

Case Study

Richard Stockton College
Pomona, NJ
Founded 1971
Undergraduate enrollment: 6,766

Richard Stockton College, a mid-sized public liberal arts college, is home to a significant sustainability feature. It features one of the world's largest single closed loop geothermal HVAC (heating, venting, and air-conditioning) systems which amounts to 1,741 tons of installed geothermal heating/cooling capacity. It was installed in the 1990s, and by virtue of its large scale, it serves as a model for other colleges pursuing this innovative type of renewable energy.

The sticker price on the project was significant: more than $1 million was raised in order to make the project a reality. Geothermal energy is currently an expensive, if plentiful, energy alternative. It also takes up a lot of space: more than an entire parking lot at Richard Stockton was dedicated to housing it. This is partly because of the need for 400 heat exchange wells drilled below ground about 425 feet.

Naturally, a main incentive for pursuing this costly project was to reduce fossil fuel use on campus. Supporters of the plan hoped that the new plant would reduce electricity consumption by 25 percent and natural gas consumption by 70 percent. This estimate has proven to be fairly accurate, according to administrators. Moreover, and of great importance to those seeking to replicate the system, the project reached the payback point, so that the savings associated with using geothermal energy now exceed the initial investment in the project. With today's rising fuel prices, investing in renewable energy that pays for itself is exceedingly valuable.

According to Alice Gitchell, a staff member of the Natural Sciences and Mathematics Department, the geothermal project can in part be thanked for a reported 13 percent overall carbon emissions reduction on campus. That figure is all the more impressive because the college grew significantly during that period.

In addition to impressive emissions reductions, the geothermal project helped turn the college into a renewable energy tourist attraction. Now visitors from all over the world come to campus to see this impressive technology. Staff and engineers from the college are also given the opportunity to travel and discuss what they've accomplished.

Resources:

Richard Stockton College Geothermal Project:
http://intraweb.stockton.edu/eyos/page.cfm?siteID=82&pageID=26

Activity:

Geothermal energy is one example of a new renewable energy source that remains underutilized. However, with increasing investment in research and development for the technology, the day is not far off when it will be much more common. Get in small groups and imagine your group represents your state legislature. Would you push for an increased use of alternative energy in your state? Why or why not? List several advantages and disadvantages to switching to energy alternatives. Which list is longer?

Questions:

1. Other than cost, what might be the main barriers to wider adoption of geothermal energy?

2. Knowing that resources such as coal, oil and natural gas are non-renewable, explain why the use of alternative sources of energy remains low. Do you think this will change in the near future? Explain your answer.

3. Energy conservation is often overlooked in discussions of energy alternatives. List at least 5 simple things you can do to start conserving energy today. Are you willing to do them? Why or why not?

Topic 22: Waste Management

Causes and Consequences

CAUSES AND CONSEQUENCES

Until we can achieve truly sustainable closed-loop systems, our activities will continue to generate large amounts of municipal solid waste, industrial solid waste, and hazardous waste. **Excessive waste generation** results from several causes and exerts various impacts on human health and ecological systems. Fortunately, there are many steps we can take to reduce our generation of waste.

Write in two causes of excessive waste generation in the spaces provided. Then write in two consequences (impacts on the environment, human health, or quality of life) that result. Finally, offer two solutions to this issue and its consequences. One cause, one consequence, and one solution have been filled in for you, providing examples.

CAUSES
- Rising consumption of material goods
- _____
- _____

→ Excessive Waste Generation →

CONSEQUENCES
- Groundwater contamination from landfills
- _____
- _____

SOLUTIONS
- Reuse items whenever possible instead of throwing them away
- _____
- _____

Sometimes solutions can have unintended consequences. Can you think of an undesired consequence that one solution to this issue might have? How might we then deal with *that* consequence?

Copyright © 2008 Pearson Education, Inc., publishing as Pearson Benjamin Cummings

Interpreting Graphs and Data

Using 1990 data from 149 countries, David Beede, an economist at the U.S. Department of Commerce, and David Bloom, a Professor of Economics at Columbia University, examined global patterns in the generation and management of municipal solid waste (MSW). Beede and Bloom were particularly interested in the relationships among wealth, population size, and per capita generation of MSW. Their results are presented in the table on the following page.

Income category of nation	Total MSW generation		Population size		Pounds MSW per capita per day
	Millions of tons/year	% of world total	Millions of people	% of world total	
Low	658	46.3	3,091	58.5	1.17
Low-middle	160	11.2	629	11.9	
Upper-middle	212	14.9	748	14.2	
High	393	27.6	816	15.4	
All economies	1,422	100.0	5,284	100.0	

Data from Beede, D. N. and D. E. Bloom. 1995. The economics of municipal solid waste. *World Bank Research Observer* 10: 113–150.

Copyright © 2008 Pearson Education, Inc., publishing as Pearson Benjamin Cummings

1. Create a bar chart on the next page with the four income categories of nations as entries on the *x* axis, and with percentages from 0 to 60 on the *y* axis. For each category of nation, plot as paired bars the values of MSW generation and population size as percentages of the world total.

2. Using the data for total MSW generation and for population size (and remembering that there are 2,000 pounds in a ton), calculate the pounds of MSW per capita per day for each category of nation, and enter these values in the table.

3. Now add a second *y* axis to your graph, on the right side, ranging from 0 to 3. Plot the values you calculated for per capita waste generation for each of the four income categories of nations, placing them as data points connected by a line. Describe in general terms the relationship between wealth and per capita generation of MSW. Can you offer at least one possible reason for the trend that you see?

4. Do you think it's possible for wealthy nations to reduce their per capita MSW generation to the rates of poorer nations? Why or why not?

Calculating Ecological Footprints

The 15th biennial "State of Garbage in America" survey documents the ability of U.S. residents to generate prodigious amounts of municipal solid waste (MSW). According to the survey, on a per capita basis, South Dakotans generate the least MSW (3.82 lb/day), and Indiana residents generate the most (11.37 lb/day). The average for the entire country is 7.25 lb MSW per person per day. Compare this number to the data in the "Interpreting Graphs and Data" table. Now calculate the amount of MSW generated in 1 day and in 1 year by each of the groups indicated, at each of the rates shown in the accompanying table.

Groups generating municipal solid waste	Per capita MSW generation rates							
	U.S. average (7.25 lb/day)		Indiana (11.37 lb/day)		"High-income" countries (2.64 lb/day)		World average (1.47 lb/day)	
	Day	Year	Day	Year	Day	Year	Day	Year
You	7.25	2,647						
Your class								
Your town								
Your state								
United States								
World								

Data from Simmons, P., et al. 2006. The state of garbage in America. *BioCycle* 47: 26.

1. Suppose your town of 50,000 people has just approved construction of a landfill nearby. Estimates are that it will accommodate 1 million tons of MSW. Assuming the landfill is serving only your town, for how many years will it accept waste before filling up? How much longer would a landfill of the same capacity serve a town of the same size in another industrialized ("high-income") country?

2. Why do you think U.S. residents generate so much more MSW than people in other "high-income" countries, when standards of living in those countries are comparable?

Case Study

Electronic Waste
Westerly Innovations Network
Westerly, Rhode Island

With growing computer usage worldwide, electronic waste, or e-waste, has emerged as a new challenge for waste management. Fourteen-year-old Alexander Lin, from Westerly, Rhode Island, took up the challenge of e-waste in his community and went on to receive international notoriety for his efforts.

Lin first became aware of the issue in 2004 when he read an article in the newspaper. He ended up dedicating a community service group he previously helped form, the Westerly Innovations Network (WIN), to the cause. One of WIN's first projects included a recycling drive that collected 21,000 pounds of e-waste, resulting in the creation of a permanent e-waste receptacle in Westerly. Considering that some communities still lack recycling facilities for glass or aluminum, this is quite a feat. The center has collected more than 60,000 pounds of waste. WIN helped draft and pass an ordinance in Westerly and later state legislation banning the dumping of e-waste.

Lin took the e-waste issue to his school as well, and persuaded the superintendent and director of technology to change the school's computer curriculum so as to include the refurbishing of donated computers. This resulted in the collection, restoration, and distribution of 260 computers to students in his hometown, and also in Sri Lanka and Mexico. Alex recently brought his efforts to bear building an Internet café in Cameroon.

His latest project is even more ambitious—he is setting up a pilot project for providing refurbished computers to youth around the world through the United Nations Environment Program and launching a partnership program between U.S. businesses and schools.

Alex is high school age, and has already won awards for his work, including the Brower Youth Award, given every year to six youth environmental leaders. His example is encouraging for anyone seeking to tackle an enormous environmental problem seemingly all alone. Who would have thought that starting in his hometown would bring him to working at an international level!

Resources:

Westerly Innovations Network: *http://www.w-i-n.ws/*

Brower Youth Award:
http://broweryouthawards.org/userdata_display.php?modin=50&uid=144

Environmental Protection Agency e-waste information: *www.epa.gov/e-Cycling*

Activity:

One of the ways Alex made such an impact reducing e-waste was by hosting large-scale collection days, where people throughout the community brought old computers, printers, and other tech equipment to a common location. Say you wanted to reduce the amount of e-waste thrown away on your campus by 20% in a single year. How would you collect this e-waste, and how would you educate your fellow students about what constitutes e-waste? Would you rely on a model similar to Alex, with centralized collections? How would you differentiate between students' personal computers and those owned by the school faculty and/or staff? Write your answers in the space below.

Questions:

1. Do you think that companies that produce electronic products should aid in the recycling and reuse of such products? Why or why not?

2. What are five things you waste on a daily basis? What else could you do with these items?

3. Have you ever visited a landfill? If not, do you think it would be a worthwhile means to educate people about the need for waste reduction? If yes, what did you learn?

Topic 23: Sustainable Solutions

Interpreting Graphs and Data

An undergraduate class at Pennsylvania State University conducted an ecological assessment of one of their biology laboratory buildings in response to the question: "How is this building like an ecosystem?" The result of their assessment was a 52-page report outlining ways to reduce the ecological footprint of the Mueller Laboratory Building in the areas of energy use, water use, communications/computing, furnishings/renovation, maintenance, and food. The students found ways to save an estimated $45,500 per year in the cost of energy alone for a building occupied by 123 scientists and support staff. Their data on the current use and potential savings in the energy component of the ecological footprint are shown in the graph.

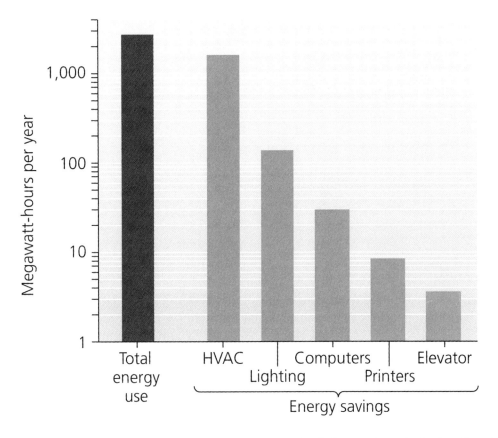

Total annual electrical energy use (red bar) and potential energy savings in five areas (orange bars) at Penn State University's Mueller Laboratory Building. The y axis is logarithmic, with each tick mark representing an increase equal to the value indicated below it on the axis. Data from Penn State Green Destiny Council. 2001. *The Mueller Report: Moving beyond sustainability indicators to sustainability action at Penn State.*

1. From the graph, estimate the amount of potential savings for each of the five areas identified in the Mueller Report. Approximately what percentage of the total electrical energy use of the building do these savings together represent?

2. What was the approximate cost of electricity in cents per kilowatt-hour used to calculate the savings of $45,500? Do you think this cost may have changed since the report was issued in 2001? How and why?

3. How is the building where you take your course like an ecosystem? How is it not like an ecosystem? Is your building operating sustainably? What improvements do you think could be made to it?

Calculating Ecological Footprints

As we have seen throughout your text, individuals can contribute to sustainable solutions for our society and our planet in many ways. Some of these involve advocating for change at high levels of government, business, or academia. But plenty of others involve the countless small choices we make in how we live our lives day to day. Where we live, what we buy, how we travel—these types of choices we each make as citizens, consumers, and human beings determine how we affect the environment and the people around us. As you know, such personal choices are summarized (crudely, but usefully) in an ecological footprint. Turn back to the "Calculating Ecological Footprints" exercise in Topic 1 of this activity book (p. 3), and recover the numerical value of your own personal ecological footprint that you calculated at the beginning of your course. Enter it in the table to the right. Now return to the same online ecological footprint calculator that you used for Topic 1's "Calculating Ecological Footprints" exercise. For many of you, this will have been *www.myfootprint.org* or *www.ecofoot.org*. Take the footprint quiz again, and calculate your current footprint.

	Footprint value (hectares per person)
World average	2.23
U.S. average	9.6
Your footprint from Chapter 1	
Your footprint now	
Your footprint with three more changes	

1. Enter your current footprint, as determined by the online calculator, in the table. How does this value compare to your footprint at the beginning of your course? By what percentage did your footprint decrease or increase? If it changed, why do you think it changed? What changes have you made in your lifestyle since beginning this course that influence your environmental impact?

2. How does your personal footprint compare to the average footprint of a U.S. resident? How does it compare to that of the average person in the world? What do you think would be an admirable yet realistic goal for you to set as a target value for your own footprint?

3. Now think of three changes in your lifestyle that would lower your footprint. These should be changes that you would like to make and that you believe you could reasonably make. Take the footprint quiz again, incorporating these three changes. Enter the resulting footprint in the table.

4. Now set as a goal reducing your footprint by 25%, and experiment by changing various answers in your footprint quiz. What changes would allow you to attain a 25% reduction in your footprint? What changes would be needed to reduce your footprint to the hypothetical target value you set in Question 2? Are you willing to make these changes? Why or why not?

Case Study

The Campus Climate Challenge: A Campaign of the Energy Action Coalition

You've read about sustainability programs at many colleges and universities. At times, it can seem overwhelming to keep all the stories straight, and also a bit difficult to understand how the stories fit together. To fill this need, in 2004 a small group of young people formed the Energy Action Coalition as an effort to aggregate the efforts of student and youth activists working to green their campuses. Specifically, they focused on student initiatives to encourage the use of clean energy on their campuses and in their communities. Since its inception in the mid-2000s, the Energy Action Coalition has grown to over 40 organizations and is active in all 50 states.

The main program of the Energy Action Coalition is the Campus Climate Challenge, an initiative involving the different member organizations of the Energy Action Coalition to encourage the students and youth they work with to win clean energy victories on their campuses. This can mean anything from students passing a campus fee to help finance the purchase of wind power, to a university committing to carbon neutrality, to a college president signing onto a clean energy pledge.

So far, some key victories have occurred throughout the University of California system, at the University of Tennessee-Knoxville, and at New York University. Over 700 campuses have signed on to be a part of the Campus Climate Challenge, and many are forming campus organizations specifically dedicated to fighting climate change.

One of the ways in which students participate in the Campus Climate Challenge and green their schools is by "power-mapping" their campuses. This involves making a chart of how key decisions are made on campus, who makes them, and how those decision-makers are influenced. These power maps are then used to build the structure of a campaign. For example, if the key decision-maker is the college president, a power map will then determine who in turn influences the president (faculty? the Board of Trustees?), and further on down to the student level. From there, students determine where to apply pressure to advocate for a certain decision. This tool proves very effective and is used by countless student organizations.

Resources:

Campus Climate Challenge: *www.climatechallenge.org*

Energy Action Coalition: *www.energyaction.net*

Activity:

Organize into small groups and discuss the following questions: Would you support a $5 per student "green fee" to be used to pay for sustainability initiatives on your campus? Why or why not? What are some things you would like to see your campus do to become more sustainable? Why do you think these steps are not currently being taken? What could you do to help make them a reality?

Questions:

1. Do you think students can influence your college or university to tackle the issue of sustainability? Explain your answer.

2. Besides colleges and universities, what other institutions should see advantages in instituting large-scale sustainability initiatives?

3. What are five sustainability choices you could make in your own life that you are not making currently? What would cause you to start pursuing these choices?

Topic 24: Geology, Minerals, and Mining

Calculating Ecological Footprints

As you saw in your text, some metals are in such limited supply that they could be available to us for only a few more decades. The number of years of availability depends on a number of factors, however: On the one hand, if new deposits are discovered and are economical to extract, they can extend availability; and if recycling technologies and efforts are improved, they also will extend availability. On the other hand, if our consumption of metals increases, then this will decrease the number of years we have left to use them. Currently the United States consumes metals at a much higher per-person rate than the world does as a whole. If one goal of humanity is to lift the rest of the world up to U.S. living standards, then this will sharply increase pressures on mineral supplies. The chart below shows currently known global reserves for several metals, together with the amount used per year (each figure in thousands of metric tons). For each metal, calculate and enter in the fourth column the years of supply left by dividing the reserves by the amount used. The fifth column shows the amount that the world would use if everyone in the world consumed the metal at the rate that Americans do. Now calculate the years of supply left for each metal if the world were to consume the metals at the U.S. rate, and enter these values in the sixth column.

Metal	Known world reserves	Amount used per year	Years of supply left	Amount used per year if everyone consumed at U.S. rate	Years of supply left at U.S. consumption rate
Manganese	5,200,000	11,600		20,020	
Titanium	1,500,000	6,100		31,900	
Nickel	150,000	1,660		5,082	
Tin	11,000	300		1,290	
Tungsten	6,300	89.6		316.8	
Antimony	4,300	135		503.8	
Indium	16	0.5		2.1	

Data are for 2007, from 2008 U.S. Geological Survey Mineral Commodity Summaries. World consumption data are assumed to be equal to world production data. World reserves include amounts known to exist, whether or not they are presently economically extractable.

1. Which of these seven metals will last the longest at present rates of global consumption? Which of these seven metals will be depleted fastest at present rates of global consumption?

2. If the average citizen of the world consumed metals at the rate that the average U.S. citizen does, which of these seven metals would last the longest? Which would be depleted fastest?

3. In this chart, our calculations of years of supply left do not factor in population growth. How do you think population growth will affect these numbers?

4. Describe two general ways that we could increase the years of supply left for these metals. What do you think it will take to accomplish this?

Case Study

Defenders of the Black Hills

Due to increased concern over carbon dioxide emissions from the combustion of fossil fuels for electricity, nuclear power once again has come into the spotlight. This raises many health and environmental questions in itself, but one important part of the discussion is mining for nuclear source material. Uranium mining has a host of impacts all its own.

Many Native American reservations are home to uranium deposits. Mining for these deposits has in some cases polluted drinking water on the reservations. Because water is considered sacred to many indigenous peoples, uranium contamination of water is viewed as an especially egregious offense.

Some tribes are standing up in opposition to uranium mining. One tribal organization, the Defenders of the Black Hills, considers protecting the environment of the Black Hills (in South Dakota) as part of its mission. For millennia, the Black Hills have been considered sacred to many Native American nations in the United States and Canada, in part because the area contains what are considered by these indigenous peoples to be the oldest mountains in the world. The Defenders of the Black Hills works to fight existing and planned uranium mining and other perceived environmental injustices.

On August 7, 2007, the Oglala Sioux Tribal Council passed an ordinance declaring Pine Ridge Indian Reservation and the aboriginal territorial boundaries of the Oglala Sioux Tribe to be a "Nuclear Free Area." Later that year, Oglala Sioux Tribal Court Judge Lisa Adams ruled against Native American Energy Group and ordered this company to be removed and excluded from the Reservation for doing business without a license.

Many such battles over mining occur in courtrooms and at public hearings, but the consequences affect thousands of people living near the mines. Defenders of the Black Hills is one of many groups that has risen to the challenge of engaging more people in this struggle through education and advocacy.

Resources:

Defenders of the Black Hills: *www.defendblackhills.org/joomla/index.php*

Uranium is Not My Friend: www.*uraniumisnotmyfriend.org/pine_ridge*

Western Mining Action Network: *www.wman-info.org*

Activity:

Organize into small groups and imagine your group is a company that mines uranium and you have located a potential new location to mine. How would you convince local people and officials that your company should be allowed to mine in the area? Do you think you would face substantial resistance? Why or why not?

Questions:

1. Explain why uranium mining continues on some Native American reservations. Do you believe this should continue? Why or why not?

2. Under the General Mining Act of 1872, mining companies can take control of public lands that have known resources, including uranium, at prices that haven't changed since the passing of the law. Often, these prices can be less than $5.00 an acre. What do you think of this practice? Why do you think this law has not been changed to reflect current land prices?

3. Would you like to see a labeling system whereby products' labels tell you where their raw ingredients were mined? What effects might such a system have? Explain your answers.